Shatterproof

ALSO BY TASHA EURICH

Insight
Bankable Leadership

Shatterproof

How to Thrive in a World of Constant Chaos
(And why resilience alone isn't enough)

Tasha Eurich

Little, Brown Spark
New York   Boston   London

Copyright © 2025 by Tasha Eurich

Hachette Book Group supports the right to free expression and the value of copyright. The purpose of copyright is to encourage writers and artists to produce the creative works that enrich our culture.

The scanning, uploading, and distribution of this book without permission is a theft of the author's intellectual property. If you would like permission to use material from the book (other than for review purposes), please contact permissions@hbgusa.com. Thank you for your support of the author's rights.

Little, Brown Spark
Hachette Book Group
1290 Avenue of the Americas, New York, NY 10104
littlebrownspark.com

First Edition: April 2025

Little, Brown Spark is an imprint of Little, Brown and Company, a division of Hachette Book Group, Inc. The Little, Brown Spark name and logo are trademarks of Hachette Book Group, Inc.

The publisher is not responsible for websites (or their content) that are not owned by the publisher.

The Hachette Speakers Bureau provides a wide range of authors for speaking events. To find out more, go to hachettespeakersbureau.com or email hachettespeakers@hbgusa.com.

Little, Brown and Company books may be purchased in bulk for business, educational, or promotional use. For information, please contact your local bookseller or the Hachette Book Group Special Markets Department at special.markets@hbgusa.com.

ISBN 9780316566551
LCCN 2024945378

Printing 1, 2025

LSC-C

Printed in the United States of America

To my chosen family—most especially Sarah Gibson Daly (my dearest friend), Marshall Goldsmith (my honorary dad), and Alan Mulally (my fiercest supporter)

"The world breaks everyone, and afterward many are strong at the broken places."

—Ernest Hemingway, *A Farewell to Arms*

Contents

Introduction: One Setback Away from Breaking   3

Part One
THE SURPRISING SCIENCE OF TWENTY-FIRST-CENTURY THRIVING

1: Welcome to the Chaos Era   15
2: The Three Myths of Resilience   26
3: Hitting Our Resilience Ceiling   43
4: The Second Skill Set   54

Part Two
THE SHATTERPROOF ROAD MAP

5: Step 1: Probe Your Pain   71
6: Step 2: Trace Your Triggers   89
7: Step 3: Spot Your Shadows   106
8: Step 4: Pick Your Pivots   118

Part Three
SHATTERPROOF TRANSFORMATIONS AND TOOLS

9: Crafting Confidence   137
10: Crafting Choice   159
11: Crafting Connection   181

Contents

Conclusion: Building a Shatterproof Life	203
Epilogue	214
Acknowledgments	219
Appendices	223
Notes	249
Index	315

Downloadable bonus chapters
    Please visit **www.shatterproof-book/bonus**
    to download the following bonus chapters:
    Building Shatterproof Teams
    Building Shatterproof Organizations

Shatterproof

Introduction

One Setback Away from Breaking

Ever-resilient Emily will never forget the day she collapsed under the weight of a feather.

One chilly February morning, Emily was completing a routine kindergarten drop-off before heading to work. As the wind stung her cheeks, she wearily hoisted five-year-old Clark from her Subaru. Her mind was racing faster than her footsteps, filled with the day's overflowing to-do list and a growing concern for her son's well-being.

Clark had been different lately: sullen, defiant, uncooperative. At home, tantrums erupted frequently and without warning; at school, his teachers were increasingly worried about his disruptive behavior. Earlier that week, Clark had tearfully confessed to being bullied by an older girl in his class, leaving Emily and her husband with nothing to do but stew as they waited for a meeting with the principal the following Monday.

Then, as Emily and Clark hurried across the school parking lot, her dejected son pointed out a child walking with her mother. "That's her, Mama. That's Greta. The one who picks on me."

No one who knew Emily, including Emily herself, could have predicted what happened next.

"I . . . saw . . . red. I marched over, pulling Clark behind me by his little hand and I—honestly, I lost my mind." Her voice trembling with fury, she confronted the little girl. "You!" she pointed an accusatory finger, "You need to stop picking on my son!"

Greta's mom instinctively stepped in front of her daughter, demanding that Emily address the issue through proper channels. Instead, Emily impulsively seized a bag of school supplies from the woman's grip. Soon the two mothers were in a tug-of-war over glue sticks and glitter as their bewildered children looked on.

The altercation was eventually broken up by two parents, and Greta's mom threatened to call the police. "Thank God no one had their phone out," Emily said. "If that went online, I'd lose my job . . . at best. And just *imagine* how this will look to the principal. I'm the crazy woman who raised a crazy child."

The truth is, Emily was *anything* but crazy.

On the contrary, I knew my friend to be a master juggler of life's endless demands: a loving wife and mother of two young children who seamlessly balanced her family life with her taxing role as a nurse practitioner at a local hospital, all while constantly showing up for her extended family and friends. She'd even started a weekend business selling beauty products to boost her kids' college funds. Emily had always faced life's challenges with unwavering resolve and positivity, including during her husband's recent health crisis. But that winter morning, she finally snapped.

As Emily told me about the Parking Lot Brawl, I learned that my seemingly cool and collected friend had felt the cracks forming for some time. Stress was materializing more often, and hitting harder, across multiple areas of her life. At work, constant pressure to "do more with less" left her falling farther and farther behind as she struggled to keep up with an ever-expanding to-do list. Like many high achievers, the thought of asking for help made Emily queasy—not just because it signaled that she couldn't handle everything but also because she knew that everyone else was just as swamped as she was.

Clark's issues weighed particularly heavily on her. Most nights, Emily lay awake watching the red numbers on her alarm clock tick upward, wondering where she'd gone wrong as a parent. Months earlier, when her husband initially raised concerns about Clark's behavior, Emily was already in a nearly constant overwhelmed state. At a loss for how she could handle one more problem, she'd suggested a wait-and-see approach—a response she now regretted at a cellular level.

To the outside observer, Emily had appeared to have everything under control. But on the inside, she was grappling with a growing sense of anxiety and self-doubt, terrified she would let everyone down. Of late, she found herself silently sobbing in her car more often than she cared to admit, even to herself. As to the cause, it would be easy to chalk up Emily's situation to weak coping skills, especially if you believe in the power of resilience. After all, for decades, we've been taught that no matter what life throws at us, if we can just manage to tough it out, we'll emerge stronger on the other side.

Yet when I dug into how my friend was coping with this stressful period, I discovered that she was doing almost everything right. She exercised regularly and engaged in self-care. She maintained a positive attitude, practiced gratitude, and worked to reframe challenges as opportunities. She turned to her husband and friends for social support. She'd even added a daily ten-minute meditation routine before work *and* kicked all added sugar (including her beloved gummy bears). By all objective measures, Emily should have been a paragon of mental health amid adversity. But these tried-and-true coping behaviors weren't getting to the root cause of her suffering. Instead, they were masking the symptoms, like a thin layer of paint applied to cover cracked concrete. And just below the surface, those cracks continued to widen.

Emily's commitment to pretending she was okay was so strong and so effective that virtually no one knew anything was wrong. Anyway, who was she to complain? She was just going through a difficult spell, like everyone does at one time or another. Eventually, she told herself, the playground bully would tire of picking on her son, things at work would start to normalize, and her husband would feel well enough to offer more help and support around the house. As she kept reminding herself, *God will never give me more than I can handle.*

But to Emily's great frustration, her mental health kept spiraling. Every day was a marathon of anxiety and dread. All she could think was, *I've always been such a tough person—so why can't I handle this?*

Introduction

That's when she called me.

"You've spent twenty years coaching some of the world's most powerful CEOs [chief executive officers] and executives, and the stakes are too high for *them* to fall apart." Then, in an uncharacteristically frantic tone, she pleaded, "You have to tell me why none of my coping tools are working and what I'm supposed to do!" I thought of a recent conversation with a coaching client, a CEO who was leading a massive business transformation. He admitted, "I thought I was doing fine with the unrelenting change. Then one day, on a call with my team, I started screaming at them. So . . . I guess I'm *not* fine."

Suddenly, I realized that neither Emily nor my client were outliers.

A few years back, I'd published my second book (*Insight*) on the link between self-awareness and success. It took me all over the world to speak to people from every walk of life. Along the way, I noticed that clients, audiences, and readers alike were increasingly asking variations of one question: "How do I handle all this chaos endlessly swirling around me?" While few were strangers to setbacks and stress, navigating them seemed to be getting harder. This hinted that our current solutions weren't working the way they were supposed to.

Here's something to know about me: with a PhD in a highly quantitative field of psychology,[1] I think "research says" are probably the two most exciting words in the English language. (I've even spent the last ten years conducting empirical research on self-awareness just for fun, without even being employed by a university requiring me to do so.) Therefore, in Emily's cry for help, I saw a new way to deploy my nerdery.

So (obviously), I assembled a research team and hatched a comprehensive program to push past the old platitudes about how to navigate stress and setbacks to find out what really worked. Our investigation involved synthesizing more than 1,200 scientific articles, surveying thousands of people across a half dozen data sets, and analyzing more than

---

1. Industrial/organizational psychology, to be exact. My field, among other things, invented personnel assessment and selection for soldiers during World War I. Now, we apply behavioral science to the workplace to improve performance, individual well-being, and organizational effectiveness.

three hundred in-depth interviews with a global sample of working adults. Our research questions mirrored Emily's: Why were existing coping strategies no longer working, and was there a better way to shore up our mental health and well-being? Fundamentally, we wanted to learn how humans can keep from shattering in a world that seems intent on trying to break us—and occasionally succeeds.

I wish I could tell you these answers came easily. The truth is that it took nearly five years to confidently uncover them.[2] And once we did, our findings upended much of the conventional wisdom about how to thrive in the face of adversity.

I wrote *Shatterproof* for my fellow **stressed-out strivers**:[3] goal-oriented people seeking success and fulfillment, who feel exhausted by chronic, compounding challenges across multiple areas of life (work, career, romantic relationships, family, friends, health, community, and the world). Because Emily is far from alone in feeling run down by the unrelenting demands of life. How often do you feel like you're moving at full speed but unable to catch up? Or pretend you're fine when you're not? Or fear that you're one crisis away from shattering?

We are working more than ever but feel like we're never doing enough. We're disconnected and exhausted but too afraid to ask for support or advocate for our needs. We strive to appear "fine" on the outside, but on the inside feel crushed by fear, anxiety, and self-doubt. Then, when we break, we blame ourselves for not being tough enough.

As I mentioned earlier, we've long been taught that by strengthening our resilience muscles, we not only can survive anything, but grow stronger in the process. However, current research (mine and others) casts doubt on much of this "grit and bear it" gospel.[4] And despite the literature showing the power of resilience, many of the most resilient among us are still struggling.

---

2. My physical therapist, Kendra, whom I have seen weekly for fifteen years, dubbed this research project the singular bane of her existence due to its lack of a speedy and satisfying resolution.
3. **Throughout this book, I'll bold key terms and phrases.**
4. A slight variation on the typical phrase "grin and bear it."

Introduction

In this book, I will argue two things. First, because resilience is a limited resource, it alone may no longer be a complete coping strategy in this increasingly chaotic world. And second, the best response to constant chaos is not merely to survive it, but to harness it in order to become the best version of ourselves. In the pages ahead, you'll learn a new set of scientifically supported strategies for doing just that, so you can feel more energized, confident, and ready to face future challenges—from small but recurring setbacks to major life-defining crises.

Yet as we trudge through each day trying to keep our heads above water, accessing the best version of ourselves might feel like a distant fantasy. Indeed, with change as our constant companion, we will inevitably bend or even break under the weight of life's stressors, sometimes more often than we'd like. But it doesn't have to be this way. As we'll see throughout this book, **in the times that break us, we can also uniquely remake us—and this is what it means to become shatterproof.**

By following the four steps of the Shatterproof Road Map in the chapters ahead, you will learn to go beyond merely bouncing back from stress or setbacks and begin to confidently channel them into forward growth. Part empowering manifesto, part scientific exploration, and part how-to guide, I hope *Shatterproof* will provide you new clarity and language around a few experiences you may have struggled to explain or describe, while setting you on the path to feeling better, doing better, and living better than you ever have before.

This book is divided into three sections. Part One reveals the surprising science of thriving in the twenty-first century: what drives it, what doesn't, and what gets in the way. Chapter 1 will confirm your hunch that everyone *is* especially exhausted right now. You'll discover the three evolutionary design flaws that make perpetual change and uncertainty so challenging for humans, so you can better understand and manage your reactions to the stress of everyday life. Chapter 2 reveals the surprising limits of resilience and presents the three biggest myths about what it can, and can't, do for us. You will also learn about **grit gaslighting:** the insidious way that we, and others, perpetuate the mistaken belief that quiet endurance is the only "right" way to navigate tough times.

Chapter 3 explores another phenomenon you might have experienced—the phenomenon of **hitting your resilience ceiling.** You'll see what happens when your resilience runs out, how to know when you're close to your limit, and why it's time for a new, more sustainable approach to cope with constant change. Finally, Chapter 4 introduces a **"second skill set" to complement the first skill set of resilience**—one that doesn't just keep us productive and sane in the face of ongoing challenges but also helps us navigate future ones with calm confidence.

In Part Two, we discover how the four steps of the **Shatterproof Road Map** present a unique pathway for growing through tough times. Step 1, discussed in Chapter 5, is facing and tracing our pain. After learning why we often overlook or suppress negative emotions, and the detrimental impact of doing so on our well-being, you'll discover how to see your pain as a power source rather than a personal failure, plus several practical tools to overcome avoidance. Step 2, covered in Chapter 6, is to identify the situational triggers that signify our unmet psychological needs. You'll see why reacting to triggers doesn't make us weak—instead, it signals that our needs for confidence, choice, or connection are being thwarted—and you will discover how to regain control of your reactions and find new routes to need fulfillment.

Chapter 7 explores why chronic stress turns us into a person we barely recognize. With step 3 of the Shatterproof Road Map, you'll uncover your shadow goals and habits—automatic yet self-limiting responses to need frustration—and learn to harness that insight to guide behavioral change. Finally, Chapter 8 outlines step 4 of becoming shatterproof: picking your pivots. By consciously turning to new goals and habits that are more aligned with your needs, you'll be positioned to significantly reduce your levels of stress, overcome your greatest challenges, and achieve your highest ambitions—all without hitting your resilience ceiling.

Introduction

Part Three is a deep dive into the three primary pivots—based on our three fundamental human needs that are scientifically shown to deepen our happiness and well-being. Chapter 9 reveals how to restore **confidence** by rebuilding your self-worth in the face of self-doubt. Chapter 10 addresses how to reclaim **choice,** empowering you to lead a values-driven life with minimal pressure and maximum authenticity. Chapter 11 helps you deepen **connection** by boosting your sense of belonging, forging mutually supportive relationships, and finding meaning in something greater than yourself.

Ultimately, we'll discover that the shatterproof journey is never a one-and-done endeavor; what's more, **becoming shatterproof doesn't mean never breaking; it means continually choosing to build back better than ever.** (Don't worry, I will reveal several secrets to sustain shatterproof habits over time.) Throughout the book, you'll find exercises, self-assessments, and practical tools to help you go beyond resilience to build a shatterproof life.

As this Chinese proverb beautifully explains, "When the winds of change rage, some build shelters while others build windmills." When you follow the Shatterproof Road Map, you will learn to better metabolize stress and reduce reactivity. You will become more focused, productive, and ready for the future. You will banish self-doubt while maximizing your mental health and well-being. You will move from "I'm one setback away from breaking" to "I'm dealing with a lot, but I've got this." And perhaps most importantly, you'll find new ways to satisfy your fundamental psychological needs, ultimately paving the way for a more authentic and fulfilling life. And after walking this very path myself—not just as a researcher, but also as a human being desperately in need of solutions—I can attest to its life-changing impact.

Before we turn the page and discover how to build a beautiful life in a world of constant chaos, let me share one thing. While I wrote this book to help you better move through everything from life's small stressors to its biggest crises, it isn't a substitute for one-on-one professional support. This is especially true if you're in crisis (not being physically safe, experiencing extreme anxiety, thinking of hurting yourself or others,

Introduction

etc.) or experiencing symptoms of trauma (flashbacks, persistent negative or intrusive thoughts, difficulty concentrating, sleeping, eating, etc.). In that case, please put this book down and seek professional assistance immediately. Appendix A contains several free, 24-7 resources. And please remember, as the great Broadway composer Stephen Sondheim once penned, "No one is alone. Truly. You are not alone."

Now, let's dive in, shall we?

PART ONE

The Surprising Science of Twenty-First-Century Thriving

Chapter 1

Welcome to the Chaos Era

> "The hardest thing in this world is to live in it."
> —Buffy the Vampire Slayer

In the vast Canadian wilderness, where lush forests once stretched as far as the eye could see, a tiny creature the size of a thumbtack was unleashing a colossal wave of destruction. It was the early 1970s, and the spruce budworm was relentlessly devouring vast swaths of northern Canada's balsam fir and spruce trees.

For two and a half centuries, these winged invaders descended like clockwork every thirty to forty years. Through all six previous epidemics, the forest always fought back to reclaim its vitality. But this time, its proven defenses didn't stand a chance.

What made this time so different? First, a string of mild winters and dry springs boosted budworm larvae reproduction, outpacing that of their natural predators. Worsening matters, aggressive forest management and fire suppression unintentionally created near-perfect conditions for total budworm domination.

Soon, tens of thousands of winged creatures were infesting virtually every spruce and fir tree over millions of acres. One observer vividly recalls a canoe trip in the affected area where endless strands of larvae rained down on him like a shower. The crisis raised wildfire risks, devastated wildlife habitats, and flatlined forest-reliant communities. In Quebec alone, the disaster caused twenty years' worth of wood supply losses. Ecologists named it Epidemic 7, and foresters dubbed it "the end of the world."

To fight the unprecedented challenge, a coalition was formed between federal and provincial governments and private landowners. And the budworm battlers knew exactly which weapon to use: the proven insecticide dichlorodiphenyltrichloroethane, or DDT. Soon, two- and four-engine aircrafts were blanketing more than 3.5 million acres with this pernicious poison.

There was just one tiny problem. The budworm battlers' big chemical offensive? It completely failed. In fact, it didn't just fail, it made the epidemic worse. (We'll learn why in Chapter 2.) As once-shimmering emerald forests became lifeless landscapes of faded gray, the devastation was shocking and total.

By 1975, Epidemic 7 became the largest spruce budworm outbreak ever recorded, spanning a stunning 136 million acres—from Ontario to Newfoundland to Nova Scotia and down half of Maine. Instead of improving the forest's resilience, the budworm battlers' game of ecological whack-a-mole was pushing it to its breaking point.

Like Canada's budworm-infested forests, **twenty-first-century humans are being pushed to the very limits of our resilience.** Years of digital disruption, geopolitical instability, natural disasters, economic volatility, and other unprecedented threats have sent shockwaves through our routines. Because the world is more connected than ever, we're more vulnerable to "phase transition events" (wide-reaching disruptions triggered by small shifts, like market crashes or social movements) and "compound extremes" (multiple co-occurring disruptions, like a natural disaster during a recession). Around the world, chaos is accelerating and uncertainty continues to rise.[1]

As professionals, parents, leaders, and citizens, we face unending demands on our time as we juggle an ever-expanding array of responsibilities.[2] At work, we strive to excel, level up, and lead while grappling with

---

1. This is the first empirical study I'm citing in the book. Because of its sheer number of citations, they will all be included as blind endnotes. (That means you won't see a symbol in the text, but I invite my fellow nerds to turn to page 249 for all the "research says" glory!)

2. A working mother recently described this experience to me as "living in a sizzling hot panini press."

growing pressure. The uncertainty, complexity, and ambiguity[3] in our environment produces even more stress and self-doubt as a dizzying array of issues constantly demand our attention. If each meeting feels more urgent and less productive than the last, you're not alone: data suggest that we're working more but accomplishing less. And, in a recent study, a staggering 75 percent of workers reported more stress now than in the previous year.

For so many of us, this chronic stress is taking a toll[4] in the form of headaches, inflammation, immune problems, cognitive issues, sleep disturbances, depression, memory loss, and more. In one study I ran with about four hundred working adults across varied jobs and industries, the majority expressed concerns about how much more stress they could handle, and many reported feeling less motivated, less engaged, and less like themselves, as though they were "emotionally coasting" or powering through on autopilot.

With so many demands on our time and attention, we're increasingly driven by "**mustivation**"—doing things because we *must* or *should*—rather than motivation. Our struggle to keep up comes with guilt, anxiety, and lingering dread about letting others down. Smartphones further blur boundaries, and balance eludes us as we juggle virtual meetings with homework help and emails with vacations. Yet despite technology's promise of greater connection, we're also feeling lonelier. Because it's a little *too* easy to stay at home in our pajamas watching Netflix and ordering from DoorDash, we're spending less time with our friends.

On a personal level, the stakes have never been higher. Consider your life right now. In a word, how would you describe it? When I ask audiences this question, I invariably hear answers like "hectic," "uncertain," "chaotic,"

---

3. With the addition of "volatility," this constellation of environmental characteristics is often referred to as "VUCA" for short (Herbert F. Barber, "Developing Strategic Leadership: The US Army War College Experience," *Journal of Management Development* 11, no. 6 [1992]: 4–12).

4. One survey showed that stress is impacting mental health for 80 percent of professionals and straining relationships with family, friends, and coworkers for almost 75 percent (Mental Health America, "2023 Workplace Wellness Research," accessed September 9, 2024, https://www.mhanational.org/2023-workplace-wellness-research#:~:text=In%202022%2C%2081%25%20of%20workers,friends%2C%20or%20co%2Dworkers).

"stressful," "overwhelming," "demanding," and "exhausting." No wonder we're tapped out or teetering on the edge, as evidenced by record-low levels of life satisfaction and record highs of anxiety and depression.

Questions linger: When all this stress keeps coming, what can I do to protect my mental health? How can I find peace and purpose when I might be one setback away from breaking? Is it even *possible* to build a beautiful life in a world of constant chaos? (You know, life's little questions.)

Of course, most of us are doing the best we can. And many of us are keeping our heads above water, at least most of the time. The problem is that while our stressors keep growing and evolving, our methods for managing them have largely remained the same. To understand why this is problematic, let's again journey back—this time, a few million years. Because even as we grapple with our present and future, we humans are largely a product of our past.

THE CHAOS CONUNDRUM

At the dawn of human existence, our ancestors were explorers, traversing unfamiliar terrains in search of sustenance and protection amid ever-looming threats. Over time, their bodies developed sophisticated systems to enhance the chance of surviving imminent peril—the fangs of a tiger, perhaps, or the footsteps of an enemy tribesman. And these systems worked (otherwise, we'd have gone extinct, and you wouldn't be reading this book).

To see how these systems work, imagine an early human ancestor. For fun, let's call him Stan. One day, while hunting under a dense canopy, Stan encounters a massive tiger, its teeth glistening in the dappled sunlight. Instantly, Stan's fight-or-flight response system starts kicking into gear. His amygdala signals his hypothalamus and sympathetic nervous system: "Hey, excuse me! We're in mortal danger here!" This activates an acute stress response, readying Stan's body to counter the threat—in this case, running away as fast as he can—allowing him to escape the tiger and enjoy a peaceful evening by the campfire.

Now, fast-forward a few million years.

Anne's life is a chaotic circus of constant demands. She's always been a high achiever, first at school and then in her career. In her financial services job, despite the long hours, fuzzy expectations, and changing priorities, she takes pride in her performance. And she loves her new side hustle of public speaking, exhausting as it is to zigzag across the country multiple times each month. Then there's her third job as a loving mother to two teenagers, supportive wife to her husband of twenty-eight years, and devoted caretaker to her eighty-eight-year-old father.

Despite constantly running on adrenaline, Anne is doing just fine. Sure, her schedule is hectic, and juggling competing demands is stressful. But she has always powered through, doing what needs to be done—and as a result, she's been rewarded with a pretty nice life.

Then, a few surprises pierce Anne's tough veneer. First, after routine tests, she receives some concerning labs and a wonky colonoscopy. Both are false alarms in the end, but getting a clean bill of health involves a lot of doctor's appointments she really doesn't have time for. Anne begins to feel like she is hitting some kind of ceiling in her ability to cope; she's exercising less, sleeping less, and drinking more than usual.

Then, the kicker. On her way home from work one day, another driver rear-ends her. Even though it was a minor fender bender, a switch flips. Overwhelmed and unable to process all the emotions she is feeling, Anne's amygdala issues the same urgent warning as Stan's: *run for your life*! But with no escape route to be found, Anne simply shuts down, like an overheated iPhone. At work, she is disconnected and disengaged, missing details she normally would be on top of. At home, she's going through the motions but not keeping up. It feels like being hit by a huge wave, scattering her broken pieces all over the sand.

This experience might be familiar: a high but mostly manageable baseline of stress, punctuated by unexpected setbacks that, at some point, can become too much. And there's a reason for that. **While the threats humans face have evolved, our survival systems have not.** The same hardwired response that saved Stan from death by tiger turns out to be woefully ill-designed for human flourishing in today's world. Let's look at three ways in which this is the case.

Design Flaw 1: Bad Is Bigger Than Good

If you have a green thumb (unlike me, a notorious houseplant assassin), you've observed what's known as the heliotropic effect. Place a plant on a sunny windowsill, and over time, its leaves eagerly reach toward the sun. Just as plants crave life-giving light, humans seek positivity; it's why we're drawn to upbeat people, why we love romantic comedies, and why compliments help us connect with others. The fascinating twist: despite our preference for the positive, our brains are wired to obsess over the negative.

Consider, for a moment, the worst feedback you've ever received—now, think of the best. Chances are, the worst stands out far more than the best. Here's why: for early humans, ignoring negative signals (like the signs of an imminent bear attack) carried a much higher penalty than overlooking good ones (like finding a warm cave to sleep in). Noticing and responding to bad things kept our ancestors alive, which is why our brains are biased to see bad as *bigger* than good.

This **bad things bias** explains why we can remember four times more bad experiences than good ones; why the pain of losing money is greater than the joy of winning; why we dwell on negative information more than positive information; why the joy of good days fades faster than the pain of bad ones; and why a solitary traumatic event can leave a permanent mark. Even for the most positive people, bad things cast longer and darker shadows than we realize. That's because, as resilience expert Dr. Lucy Hone explains in her excellent TEDx talk, "negative [experiences] stick to us like Velcro, whereas positive [ones] . . . bounce off like Teflon."

In Stan's day, bad things were temporary and infrequent. Today, bad things are unpredictable, overwhelming, and demand constant attention: be they minor inconveniences like traffic jams and canceled flights, moderate challenges like missed deadlines and rising costs, or existential crises like climate change, death, and war. Without intervention, bad things bias doesn't protect us from danger; it plunges us into a chronic state of stress.

Design Flaw 2: The Cortisol Conundrum

Living in perpetual fight-or-flight mode isn't just stressful, *it drains the very resources we need to cope with stress.*[5] To see why, let's return to Stan's tiger encounter. As his body registers the threat, his adrenal glands activate, releasing two key stress hormones. Adrenaline gears Stan's body up to fight or flee—widening his pupils for clearer vision, tensing his muscles for quicker responses, spiking his heart rate and oxygen for superstrength. Noradrenaline sharpens his attention, boosts his strength, and redirects energy away from nonessential cognitive activities like deep thinking, and toward his heart, lungs, and muscles. Before long, a third hormone, cortisol, joins the party, providing power to keep pushing until he reaches safety.

A few million years later, Anne's stress response system operates identically—though you've probably noticed a difference in the threats she faces. While Stan's tiger signaled mortal danger, **Anne's stressors are more abstract and psychological**—a demanding job and side hustle, her "third job" as a caretaker, a health scare, a fender bender. But because her prehistoric body doesn't distinguish from these things and a tiger, each causes her cortisol to spike, making her more and more anxious and reactive. Taken together, this cocktail of stress hormones impairs the clear thinking, communication, and control that successfully managing modern-day threats requires.

Another difference? Stan's stressors were temporary and infrequent, while **Anne's stressors are chronic, cumulative, and extend across**

---

5. In fact, over time, our stress system can become so exhausted that it stops producing cortisol. Helpful as this sounds, cortisol is essential for daily functioning—otherwise, we'd have no energy to get out of bed! (Christine Heim, Ulrike Ehlert, and Dirk H. Hellhammer, "The Potential Role of Hypocortisolism in the Pathophysiology of Stress-Related Bodily Disorders," *Psychoneuroendocrinology* 25, no. 1 [2000]: 1–35).

multiple domains of her life (work, family, health, etc.). Recall that after escaping the tiger, Stan's hypothalamus shut down his stress response system; yet without an "all-clear" signal, Anne's stays awake. Indeed, while Stan was able to flee from the tiger, many modern-day daily equivalents are inescapable, like emails from our boss, causing a constant flood of cortisol that overwhelms our brains. The cortisol conundrum means that modern-day humans are almost constantly experiencing a heightened stress response, which is difficult (sometimes impossible) to switch off.

Design Flaw 3: The Anarchy of Uncertainty

The last piece of the puzzle is uncertainty. Uncertainty involves unpredictable situations, a lack of control, or missing information. When Stan was uncertain, it was safer to assume the worst (i.e., better to be stressed than dead), which is precisely why uncertainty triggers our fight-or-flight system. Living under chronic uncertainty takes a massive toll on performance and well-being. It threatens our sense of safety, increases worry, anxiety, and reactivity, and makes us desperate for answers. To cope, some people will claim certainty despite having none—which results in overconfidence about what they know, less interest in new information, greater susceptibility to conspiracy theories, and more aggression toward those who disagree with them.

Nevertheless, it's still logical to assume that a definitive bad outcome, like getting laid off, is more stressful than the *possibility* of a bad outcome, like rumors of impending layoffs. But often, this isn't the case. For example, research shows that worrying about job loss is more stressful than *actually* losing our job! If you've ever thought, "I don't even care what happens anymore, I just need to know the outcome!" you've experienced the **certainty over comfort effect.** Consider a study by neuroscientist Archy de Berker and colleagues, who found that when we see a 50/50 chance of something bad happening (in participants' case, it was getting an electric shock), our stress, reactivity, and agitation spike *higher* than when we're 100 percent sure. This might be why some psychologists see uncertainty as possibly *the* fundamental human fear.

In summary, humans' three hardwired stress responses—bad things bias, the cortisol conundrum, and the anarchy of uncertainty—simply haven't evolved to handle the complex challenges of contemporary life. Yet despite their shortcomings, they still serve an important purpose: **stressors are a signal that something isn't quite right in our lives.** Therefore, instead of ignoring or resisting these natural responses, we can choose to be curious—treating them as clues to uncover our true sources of stress, and catalysts for us to change course.

However, my research has revealed that the path to transforming stressors into strengths is very, very different from what we've been taught.

ONE STUDY TO SAVE US ALL

It was January 2020. Between keynotes and coaching commitments, I was embarking on an exciting new research program to understand how **resilience—broadly defined as the capacity to bounce back from hard things**—could keep us from breaking in a world of constant chaos. I gave it the unintentionally prophetic name "When Bad Things Happen."

By early March, I'd trained a team of ten research assistants to interview hundreds of working adults worldwide. Through a newfangled (to us) platform called Zoom, each participant shared a "bad thing" they'd experienced and then answered a series of standardized questions on how they had responded and how things turned out. The situations they described ran the gamut, from common frustrations (unfair bosses, micromanagement, criticism) to life-altering crises (betrayals, firings, fallings out)—some were sudden (like accidents), while others were slow-burning (like toxic relationships).

Initially, months of data analysis revealed only two patterns. First, "bad things" were indeed quite bad, leading to a loss of self-confidence, energy, and motivation; persistent negative emotions like sadness, anger, and anxiety; and poor performance and lower life satisfaction. For instance, during an ongoing rift with his business partner, Greg noticed he "had less to give as a family man," recalling that "I wasn't as gracious with my wife or patient with my kids . . . the issue consumed my bandwidth and affected them too." When Cara was being bullied, she was moody and unhappy and "forgot how to have fun."[6]

The second pattern was that people responded to bad things in one of two ways: they either found a way to bounce back from them—or not. (This staggering revelation would hardly get me shortlisted for next year's Nobel.) But when I dug even deeper into the data, I eventually discovered a smaller, third group of inspiring people for whom the bad thing had become a force for good in their lives. As the challenge unfolded, and after it was over, they became less stressed, more empowered, and more in control. Professionally, they felt more effective and purposeful. In their relationships, they experienced deeper connection and belonging. This group didn't just bounce back from adversity—they grew forward.

Our interviews therefore revealed three distinct outcomes of bad things: they can **break us,** they can allow us to **bounce back** to our baseline, or they can inspire us to **become better.** I became fascinated with this third group. What was their secret? Through the fall and into the winter, I spent hundreds of hours listening to each interview multiple times, trying to determine what group three was doing differently from everyone else. When I still couldn't spot any patterns, I tried to analyze the data another way, by translating interview responses into numbers. When that didn't work, I collected quantitative survey data. And more data. And I found . . . nothing.

It would take another year, numerous quantitative replication studies, and several false starts to find the answer. As it turns out, that answer wasn't all that different from the reason the budworm battlers failed, or why modern-day humans are more stressed than our hunter-gatherer ancestors. Generally, the first two groups—those who broke and those who bounced back—were applying old tools to solve new problems; tools that, like DDT and humans' fight-or-flight response, were fundamentally ill-suited for the challenge at hand. The third group had a dramatically

---

6. These psychological challenges also coincided with physical symptoms like headaches, high blood pressure, back pain, hives, insomnia, intestinal issues, undesired weight gain or loss, panic attacks, depression, infections, autoimmune diseases, and even cancer.

different approach. What they taught us would ultimately go on to challenge everything I believed about navigating adversity, and not just as a scientist, but as a human being.

CHAPTER 1: KEY TAKEAWAYS AND TOOLS

"While the threats humans face have evolved, our survival systems have not."

- **The chaos era**: An age of increasingly chronic and compounding stress across multiple domains of life, like work, family, community, etc.
- **Stressed-out strivers**: Achievement-oriented individuals grappling with more anxiety and self-doubt than they care to admit, and who tend to have a hard time asking for help.
- **Design flaws that make constant chaos so hard for humans**
  1. **Bad things bias**: Our brains are designed to see bad things as *bigger* than good ones.
  2. **The cortisol conundrum**: Living in perpetual fight-or-flight mode isn't just stressful, it drains the very resources we need to cope with stress.
  3. **The certainty over comfort effect**: The phenomenon where the *possibility* of bad things happening is more stressful than those things actually happening.
- **"When bad things happen" research**: A program examining how not to break in a world of constant chaos. Our team found that some people are broken, most bounce back, and a select few become better.
- **Stress signals that something isn't quite right**: Instead of ignoring or resisting our natural responses, we can treat them as clues to start transforming our stressors into strengths.

CHAPTER 2

The Three Myths of Resilience

"Out of life's school of war—what doesn't kill me, makes me stronger."
—Friedrich Nietzsche (and Kelly Clarkson)

It all started with a strange sensation in my arms—an insistent, prickling pain beneath my skin. Whenever I'd settle in to write, a lightning bolt shot from my elbow to my fingertips. *Ouch!*

It was early 2021, and I was feeling otherwise positive. With pandemic restrictions loosening, I knew the world would heal and we'd all bounce back. But this weird pain in my arms. What *was* that? Not wanting to burden my doctor at such a time, I resorted to my usual solution: powering through.

Then one February afternoon, the throbbing pain finally forced me to stop work early. As I slammed my laptop shut, my eyes fell on a framed photo of Deer Creek cutting through a pink sandstone cave—below it, the text: "Perseverance: In the confrontation between the stream and the rock, the stream always wins . . . not through strength, but persistence."

That frame has graced all seven of my offices since graduate school. I'm not a motivational poster–type person, but this one is small, tasteful, and reminds me of my mom. As a working mother, and later a single one, she couldn't find a job that allowed her to spend enough time with me. So, she took matters into her own hands, founding the first US company that certified and placed nannies in the homes of working parents. The same poster hung in *her* office.

The Three Myths of Resilience

My mother has always been my resilience role model. As a child, I watched her navigate a devastating divorce, single parenthood, and the highs and lows of entrepreneurship with grace. She comes from hardy stock, like her great-grandfather Alois, a penniless German immigrant who scraped together enough money to buy a small dairy farm that he passed down to his son. Our forebears showed us that challenges could—and should—be soldiered through.

Like everyone, I'd endured my fair share. At age five, my parents' bitter divorce brought court-ordered therapy, mandated visitations, and a lingering fear that it was all my fault. School was no refuge, but it did teach me that I could survive being incessantly bullied despite my desperate attempts to fit in. As an adult, I powered through a PhD program in organizational psychology by age twenty-six, and later left a cushy corporate career to start my own company.[1]

But my defining battle has been a lifetime of painful and perplexing health problems, which I'd always summoned the strength to endure—a testament, I believed, to tried-and-true resilience practices like gratitude, optimism, and active coping. For as long as I could remember, I'd touted resilience's powers, my faith buoyed by genre-defining books like *Resilience, Grit, The Obstacle Is the Way,* and *13 Things Mentally Strong People Don't Do.*

So, this minor issue of my painful arms? I just needed to be the stream: to harness my mental strength, become bigger than the obstacle, and summon all the grit I could muster. But the damned rock refused to wear down. Soon, constant pain was coursing through my entire body. Even after a full ergonomic office makeover, I struggled to work past 3:00 p.m. because of throbbing pain at the base of my skull.

New symptoms joined the party. My vision randomly blurred, making reading impossible. My "resting" heart rate often exceeded 150 beats per minute. I was unable to remember what had happened hours, minutes, or even seconds earlier (and sometimes, the names of people I've

---

1. The Eurich Group is a boutique executive development firm that helps leaders and teams succeed when the stakes are high.

known for decades). Things were getting undeniably tougher. But this certainly wasn't the worst crisis I'd faced. I tried my best to soldier on—a "keep calm and carry on (taking Ambien)" approach, if you will. Still, I found myself wrestling with more anxiety than I could remember. Each day I felt myself inching closer to some heretofore unfamiliar breaking point.

Why weren't my best resilience tools working?

I finally decided it might be a good idea to revisit the resilience research. Maybe I was missing something. I returned to the field's pioneering work, where researchers uncovered certain "**protective factors**" (like social support, positive emotions, and grit) that differentiated children and adults who withstood, bounced back from, or adapted to adversity. Later work focused on more **learnable strengths or practices** that were thought to further shore up resilience, like optimism, gratitude, exercise, and active coping. But perhaps most interesting was one recent study where researchers found that applying several different practices was the most effective route to strengthening our resilience, in the same way that mixing up our exercise routine best strengthens our muscles. This reinforced that I was on the right path—I just had to deepen and diversify my resilience practices.[2]

Naturally, my next step was to create a Supercharged Resilience Plan in the form of a printable daily spreadsheet. Each day, I gave myself checkmarks for all the practices I'd used, like meditating, taking outdoor walks (often reaching six miles a day), dialing up unsuspecting friends (who were rather confused to hear from me), finding things to be grateful and hopeful for, eating healthier, and even taking up yoga (a practice my type A self had spent years openly mocking). Soon, I was racking up checkmarks like I used to rack up United Airlines miles. I could *feel* my resilience muscles strengthening.

Admittedly, given how much time and energy this was taking every day, it was kind of exhausting. But, no pain, no gain, right?

---

2. It's worth noting that there has been some criticism of resilience strengths and practices, specifically that most of this work is correlational in nature.

Meanwhile, I saw doctors. Lots of doctors. I cycled through specialties faster than most medical students: primary care, physiatry, orthopedics, neurology, and more. The clear solution, or so I believed, was to continue checking items off my spreadsheet, let the doctors do their jobs, and remember the famous adage that *what doesn't kill us makes us stronger*.

Little did I know that in mere months, it would become painfully clear that "what wouldn't kill me" might *actually* kill me—a discovery that would propel me headlong into the depths of resilience research, following currents of knowledge that completely reshaped my beliefs about how humans can best navigate twenty-first-century adversity.

THE FATHER OF RESILIENCE

Crawford Stanley Holling's picturesque childhood unfolded against the pristine lakes and boreal forests of 1930s northern Ontario, Canada, where he developed a deep love for the natural world, especially insects and birds. When his older sister lovingly bestowed him the nickname "Buzz," the boy was delighted, insisting that everyone use his prized moniker.

After earning bachelor's and master's degrees in zoology, followed by a PhD, Buzz joined the Canadian Forest Service in 1957 as an ecologist studying predator-prey relationships. A decade later, he transitioned to teaching at the University of British Columbia, where he would soon make a pivotal discovery that would earn him international recognition.

At the time, northern Ontario was in the early throes of the budworm outbreak you read about in Chapter 1. Buzz wanted to understand why the tried-and-true tactic of widespread spraying was failing to stem the crisis. The results of his computer models were puzzling yet undeniable: in heavily sprayed areas, tree density had increased (seemingly, a win). But all that extra foliage was acting as camouflage for these insidious predators—and birds (budworms' natural enemies) couldn't spot their prey. The budworm battle, in other words, wasn't just failing to address the weaknesses that had caused the outbreak in the first place; it was unintentionally stifling the natural forces keeping the winged invaders in check.

As a student, Buzz had eagerly studied the teachings of renowned scholars like Eugene Odum and George Perkins, who theorized that ecosystems (like forests) tend toward stability, and when disturbed, would naturally recover. But the evidence staring him in the face told a much different story. While some parts of the forests were indeed self-stabilizing, others were collapsing. Buzz didn't mince words: "Mother Nature is *not* . . . in a state of delicate balance." Ecosystems were indeed capable of naturally recovering and rebalancing, but only up to a point—and the nature of some shocks made bouncing back impossible.

Buzz's bombshell went against both government policy and ecological dogma. The realization that instability and chaos weren't just temporary interlopers, but an inherent feature of the natural world, required a paradigm shift. Contrary to conventional wisdom, ecosystems didn't automatically bounce back amid major disruption. To survive in the face of unpredictable forces, what really mattered was how well they could adapt. So instead of trying to wipe out the budworms ("spray all the trees!"), the budworm battlers needed to let the ecosystem learn to manage the disruption ("let birds do their job"). **Buzz coined the term "resilience" to describe the capacity of a system to adapt to disturbance and keep functioning.**

The broader implications of his thinking were radical and intriguing. In 1999, Buzz founded the Resilience Alliance[3] to foster multidisciplinary collaboration among scientists and practitioners studying complex adaptive systems, and soon these principles were being adopted beyond ecology, by individuals, companies, and communities grappling with the increasing pace of change. Suddenly, people everywhere were asking, *What if, instead of waiting for things to rebalance, we could strengthen our capacity to adapt?*

---

3. Later renamed the Resilience Network.

A BRIEF HISTORY OF RESILIENCE

The word *resilience* comes from the Latin *resilire*, meaning "rebound" or "spring back," and across religion and philosophy, it's an eternal virtue. Christianity, as in the tale of Job, emphasizes faith, hope, and endurance in the face of hardship. Judaism honors perseverance during persecution and exile, like in the story of Exodus. Islam teaches patience and trust in God during life's trials. Buddhism acknowledges the necessity of enduring life's inherent suffering. Hinduism advocates detachment to withstand pain, and the ancient Greek philosophy of Stoicism—currently experiencing a resurgence—encourages resolutely enduring challenges with amor fati (love of fate).

But despite resilience's long and storied place in history, psychologists didn't start studying it until 1954, when developmental psychologist Emmy Werner and her team began tracking a cohort of 698 children in Kauai. By extensively assessing and observing this cohort's development, family environment, and socioeconomic circumstances from age one to age forty, the researchers hoped to discover why some children successfully rebounded from adversity while others suffered lasting psychological harm.

In the late 1960s, Werner and her colleagues began publishing their results,[4] including the heartening finding that some children *were* uniquely able to bounce back from significant challenges like economic hardship and troubled home lives. These so-called resilient children were less likely to have criminal records, psychiatric disorders, and substance abuse problems, and later in life, they were more likely to find unexpected success in school, work, marriage, and parenting relative to others who had suffered hardship.

What set the resilient kids apart, the research found, was the presence of protective factors—traits like sociability and aspects like effective parenting and access to quality education and healthcare—that helped them adapt during hard times. Yet early research on protective factors sparked debate, with some psychologists noting the dangers of

---

4. Though Werner and her team didn't start using the term *resilience* for nearly ten years.

implying that people without certain traits couldn't overcome adversity or function well amid challenges.

In response, researchers started studying the *process* for how resilient people cope and adapt to stress, and later, strategies to increase one's resilience. Soon, self-help books like *The Resilience Factor* heralded resilience as "a crucial ingredient—perhaps *the* crucial ingredient—to a happy, healthy life." Instead of being dictated by our traits or environment, the thinking went, resilience was a skill set that anyone could develop and summon at will. Suddenly, a once-obscure scientific concept became an accessible well-being tool.

But it wasn't until the 2008 global financial crisis—arguably the dawn of the chaos era—that the concept of resilience started spreading faster than a swarm of hungry budworms. Positioned as a panacea for a range of challenges, including economic inequality, natural disasters, and risk management, resilience soon cropped up in disciplines like business, education, therapy, healthcare, engineering, infrastructure, and beyond.[5]

Around this time, mainstream conversations about resilience started veering away from Buzz Holling's emphasis on the capacity to adapt and toward the idea that anyone can bounce back from anything if they're tough enough—which was when things really started going off the rails. Take, for instance, *The Art of Resilience: Strategies for an Unbreakable Mind and Body*, in which adventurer and author Ross Edgely argues that the key to achieving impossible goals, like his record-breaking swim around Great Britain, was simply a matter of cultivating "mental strength."

But the absence of evidence for such claims hasn't quelled resilience's momentum in pop culture. Taylor Swift's "Shake It Off" is an ode to resilience; public figures from Michelle Obama and Jennifer Aniston to Elon Musk have spoken openly about its importance in their lives; and companies send employees to resilience training to cultivate the "most underrated and powerful skill . . . [needed] to be successful in life."

---

5. Per Google Ngram Viewer, which analyzed over fifteen million English-language books for the word *resilience*, the term's usage increased nearly 250 percent between 2007 and 2018.

Considering all the hype, I wondered just how far the public perception of resilience had strayed from Buzz Holling's capacity to adapt. To find out, I decided to examine exactly how everyday people saw resilience. After a survey of 324 working professionals, most definitions indeed focused on the idea of strength in adversity, like mental toughness ("being strong, tough, and unmoved in challenging circumstances"), endurance ("persevering through difficulties, not giving up"), and bouncing back ("get up after falling"). Clearly, the way many of us commonly see resilience has drifted considerably from its origins.

In a case of life imitating science, these very themes were also looming large in my own life at the time. While wrestling with our team's research, I was still keeping up with my resilience spreadsheet and doctor's appointments, certain that if I could stay mentally strong and endure, I'd be able to bounce back in no time.

The one, tiny, insignificant hiccup? This approach was not working. At all.

I was convinced it was just me: other people couldn't possibly be having nagging doubts that resilience wasn't the silver bullet that self-help authors, celebrities, and some academics were promising. So, already deep into the quantitative portion of our research, I decided to find out by adding a short, validated resilience scale to one survey of over four hundred working adults. When I analyzed the results, I was shocked to find that **resilience did not, in fact, predict whether people became better and stronger after crisis.** In fact, it didn't seem to predict much about overcoming adversity at all. From the stillness of my office, I reeled. This was a staggering, almost heretical possibility. Resilience had to be the surest strategy for dealing with life's toughest moments, and bouncing back was the best way to move forward.

But what if that wasn't the case?

As I dove deeper into the research that Buzz Holling helped catalyze, I discovered just how far the mainstream view of what resilience can do for us has strayed from the science. Specifically, three resilience myths are misleading us far more than we realize.

THE THREE MYTHS OF RESILIENCE

Myth 1: Resilience helps us become better and stronger.

Truth: Resilience helps us maintain or regain our baseline strength and well-being.

If you've read any books or articles on resilience, you might have the impression that it produces endless benefits, helping us reach impossible goals, grow and thrive through challenges, and lead a happy, healthy life.

But does it really?

According to the field's most well-respected experts, resilience's true promise typically falls short of the heart-stirring ones we've come to expect.[6] Take one study on resilience during the COVID lockdowns. Of the more than one thousand people surveyed, social support, self-esteem, and other protective factors were indeed associated with resilient outcomes. A powerful headline, surely. But when we examine how the researchers defined resilience, it wasn't thriving or even adapting through tough times. *It was simply the absence of depression and anxiety.*

And by no means is this study an outlier. Several key literature reviews suggest that resilient people find equilibrium in adversity rather than becoming better and stronger (not unlike Emmy Werner's resilient children). In my own exhaustive analysis of over two hundred peer-reviewed resilience articles—including the fifty-two most highly cited[7]—I

---

6. Dear readers, the data that I am about to present to you might be surprising. If you are interested in seeing specific evidence, I "show my work" as much as possible in the endnotes for this section. If you'd like even *more* detail, I encourage you to visit www.shatterproof-book.com/resilience.

7. The number of times peer-reviewed journal articles are cited is a good indication of their quality and influence. During the week of January 15, 2024, I used Google Scholar to search for articles on "resilience." Then, I used Scinapse as a check to ensure Google Scholar did not miss any relevant articles. From there, I analyzed the definitions of the most highly cited articles on the topic of individual resilience (excluding disciplines like ecology, supply chain, etc.). Notably, beyond the fifty-two most cited articles (those with more than two thousand citations), I found the same patterns for the twenty-eight articles cited between one thousand and two thousand times.

uncovered two common benefits of resilience, neither of which support the widespread belief that it's the secret to human flourishing.

The first and most common benefit was that, put simply, resilience keeps us from falling apart. A whopping 85 percent of the most cited articles state that resilience helps us cope, maintain psychological stability, and avoid negative psychological outcomes. In other words, **resilience averts emotional disaster but doesn't leave us feeling better than we did before.**

Second, while research shows that resilience aids in adapting to adversity and uncertainty, it typically doesn't lead to sweeping transformations. Typically, resilient individuals function "better than expected" or "better than average" compared to those facing similar experiences, but they rarely rise above their baseline, normal functioning. In only 8 percent of highly cited articles—four total—did researchers argue that resilience fosters a higher level of functioning. But three-fourths based this conclusion on anecdotal, rather than empirical, evidence. Indeed, per resilience researchers themselves, "most [of us] have set the bar at the level of the normal range . . . *because [the] goal is to understand how individuals maintain or regain [normal] functioning*" (italics mine).[8]

My own research supports this as well. Across three separate studies, resilience practices—like letting go of negative emotions, changing one's mindset, seeking social support, and so on—didn't consistently predict better functioning in the face of challenges. And strangely, resilient people were no more likely to engage in these practices than anyone else. In one sample, when participants scored higher on resilience, they were actually *less* likely to ask for help or directly deal with their problems and *more* likely to fixate on bad things they wanted to let go of. In another, having a high resilience score didn't predict the belief that bad things could be a force for growth.[9] Finally, across several samples, resilience

---

8. Michael Rutter, one of the field's foremost experts, notes that "there is a misleading implication that [resilience] requires generally superior functioning rather than relatively better functioning compared with that shown by others experiencing the same level of stress or adversity" ("Resilience as a Dynamic Concept," *Development and Psychopathology* 24, no. 2 [2012]: 335–44).

9. For you statistics nerds, it was a 0.0 correlation!

didn't predict self-reported growth or improved well-being through hard times—in fact, the people who did become better were no more resilient than anyone else!

Alongside the most respected resilience research, my team's data paint a clear picture: **the primary function of resilience is helping us maintain functioning, not improve it**. And while the stability it offers can help us survive, there's surprisingly little evidence that it makes us stronger, much less reliably helps us thrive. Put another way, drawing on resilience in hard times is like rebuilding the same fragile one-story houses that were destroyed in a flood rather than constructing more durable, elevated structures capable of withstanding future disasters.

This is where the problem arises. When we start expecting resilience to do things for us that it wasn't designed for—like transforming our lives in our toughest times—we're doomed to keep making the same mistakes over and over, leaving us vulnerable to future crises.

You might find this a bit unsettling (I did!). But let me be clear: I am *not* arguing that resilience isn't a useful and potentially powerful tool. In the short term, it can absolutely help us keep it together in the face of crises—especially ones that are sudden and wholly beyond our control, like layoffs, divorce, death, disaster, and so on. And sometimes, emerging from a crisis no worse than before is an absolutely monumental accomplishment. At the same time, we would do well to start seeing resilience as a single tool designed for a certain function instead of the whole toolbox.[10] Because even if it can temporarily keep us from falling apart in the face of challenges, resilience alone isn't always enough to build back stronger.

Myth 2: Resilience is a choice.

Truth: We can't always control our level of resilience.

There's an oft-cited metaphor that resilience is a muscle. Which is to say, the more we exercise it, the stronger it becomes. This implies that anyone

---

10. Thank you to the beta reader for pointing this out!

can choose to become more resilient simply by regularly practicing a certain set of behaviors. But does the research bear this out?

Several "studies of studies" show that resilience interventions—for example, strengthening protective practices like positive thinking and social support, or learning coping strategies like reappraisal and acceptance—can boost resilience, but only slightly.[11] Others have found that resilience-strengthening interventions don't improve resilience, *at all*. One recent study of studies with over forty-five thousand data points reported that resilience interventions didn't just fail to improve resilience—they actually harmed mental health, well-being, life satisfaction, and job satisfaction, and caused higher levels of distress for those who attended resilience training versus those who didn't![12]

Indeed, believing that resilience is a choice doesn't just put us on shaky scientific ground, it exposes us to a dangerous unintended consequence: **if mental fortitude is indeed wholly learnable, as many self-help authors argue, if we fall short, it means we just didn't *try* hard enough**. As esteemed resilience researcher Michael Ungar correctly outlines this faulty thought process, we failed because we didn't "sit long enough on [our] yoga mat . . . or . . . have enough grit to overcome [our challenges]." Ungar boldly labels this conclusion "arrogant and victim blaming." On top of creating guilt and shame, this belief also shifts our focus away from the real sources of our problems, which are often internal.

That was certainly the case for Jan. Growing up with a loving family and a charmed life, she expected parenthood to be equally blissful. Then her son Andrew was born. His constant restlessness concerned her, but at first, she figured it was just a phase. Six months later, Jan contacted their family doctor about her fears. The doctor's response: "Calm down. Andrew will be fine." She chose to trust his judgment; after all, he was the expert.

---

11. Put another way, we're less likely to move from a 5 to a 10 than from a 5 to a 6.

12. At least in the workplace, one possible explanation for why some resilience interventions can hurt us comes from Margaret, a nurse. Instead of addressing the root causes of the nursing staff's stress, staffing shortages, she explains, her company is giving her and her colleagues resilience training (John Patrick Leary, "Resilience Is the Goal of Governments and Employers Who Expect People to Endure Crisis," *Teen Vogue*, July 1, 2020, https://www.teenvogue.com/story/whats-wrong-with-focus-on-resilience).

But when Andrew's problems persisted, Jan's anxiety intensified. The more she asked the doctor for help, the more he dismissed her concerns—treating her like an irrational, overly vigilant parent. Pretty soon, Jan's friends and family joined in too. One memorable example: "Andrew is fine. *You're* the one who needs therapy."

Soon, the once-confident, badass, MBA-wielding woman with a successful career was plagued by shame and self-doubt. She began to believe the real problem wasn't her son's behavior but her inability to handle it. In reality, Jan was merely the victim of something I call **grit gaslighting,** a common phenomenon where, instead of validating our stress or distress, our commitment to coping with it is questioned. Often, grit gaslighting comes from people in positions of authority or well-meaning but unaware family and friends.

Still other times, the call is coming from "inside the house." When we have been conditioned to believe that we can get through any challenge if we just try hard enough, our internal dialogue can grit gaslight ourselves. For me, during the toughest part of my medical crisis, 2:00 a.m. was the ideal time to question my mental toughness. *So many people have it so much worse than I do*, I'd think, staring at the ceiling. *What's so wrong with me that I can't handle this?*[13] (Relative to using others' suffering as fodder for self grit gaslighting, my wise friend Nick recently quipped, "Remember, it's not the oppression Olympics.")

The truth is that **several factors outside our control make it difficult to stay resilient, especially under stress.** These factors include qualities we possess at birth, like our DNA, nervous system, personality, and temperament, as well as our early childhood experiences and later life events.[14] (For an assessment of *your* risk factors, see Appendix B.) Therefore, assuming that resilience is a choice cruelly denies the challenging

---

13. In a nod to *Insight*, at least I was asking myself "what" questions instead of "why" questions!

14. While researchers are *mostly* sure that resilient people weren't simply born that way, they're still not, like, *100 percent sure* (Christy A. Denckla et al., "Psychological Resilience: An Update on Definitions, a Critical Appraisal, and Research Recommendations," *European Journal of Psychotraumatology* 11, no. 1 [2020]: 1822064).

reality many people experience. The truth is, we're not failing at resilience; our total reliance on resilience is failing us.

Myth 3: What doesn't kill us makes us stronger.

Truth: What doesn't kill us makes being resilient even harder.

In January 1889, philosopher Friedrich Nietzsche published a book containing the famous aforementioned aphorism: "What does not kill me makes me stronger."

That same month, during a morning stroll through the Piazza Carlo Alberta in Turin, Nietzsche witnessed a cab driver violently beating a horse. Rushing toward the tortured animal and shouting incoherently, the philosopher flung his arms around its neck and burst into hysterical sobs. The police were called. After an acquaintance walked Nietzsche home, his condition worsened. Gripped by delusions, he locked himself in his room, screaming at the top of his lungs. He was soon committed to an asylum in Jena and, according to his closest friend, "never emerged again."

In other words, less than thirty days after going on the record with "what doesn't kill you makes you stronger," *Nietzsche himself disproved it.*

Yet in his belief that the mere experience of adversity strengthens resilience, he was not alone. While this view is attractive, uplifting, and provides a comforting sense of invincibility, resilience researchers note that it "lacks robust empirical evidence." In fact, the people with the most limited resilience resources are often those who have experienced the most stress in their lives. This suggests **that resilience is less an infinite source of strength and more an exhaustible resource that dwindles each time we draw from it.**

In fact, the true relationship between stress and resilience generally runs counter to Nietzsche's theory: rather than boosting resilience, ongoing stress tends to *deplete* it. Over time, this can be true even for minor stressors, and particularly for chronic challenges across more than one life domain (a.k.a., what life looks like in the chaos era). As one longitudinal study of Chinese college students between 2007 and 2020 found, increasing stressors like economic challenges and social disconnection consistently *lowered* participants' resilience levels over time.

For an apt illustration of this phenomenon, let's turn to a 1998 study by psychologist Mark Muraven and his team. After students watched a distressing movie, some were asked to fully engage with their emotions, while others were asked to suppress them (i.e., resiliently endure). Students in the second group were more fatigued and less energetic; and as their psychological energy depleted, their physical energy did as well.[15] This is kind of like how our cell phone's battery capacity diminishes the more we charge it.[16] But unlike a cell phone, we can't simply upgrade our resilience to a newer model. This is a fitting example of what happens when we follow the "resilience-as-a-muscle" metaphor to its logical conclusion. When we continue working any muscle—adding more and more weight, doing more and more repetitions—it becomes fatigued, and without rest, it will fail completely. By the same token, **the belief that more stress makes us stronger merely makes us more vulnerable.**

This "what doesn't kill us" myth can be especially harmful to members of marginalized groups. Beneath it lies the assumption that the "correct" response to inequity is quiet endurance instead of challenging flawed systems. Take author Simran Jeet Singh, who is Sikh and frequently the target of discrimination because of his ethnicity. Once, he mentioned to a friend that he had an unusually difficult week at work. Instead of compassion, his friend hit him with, "You've been through much worse. This will be easy for you." In his friend's mind, if Singh could endure racism and bigotry, work drama should be a piece of cake.

---

15. One theory for why resilience helps us is that it preserves cognitive resources—but because doing so takes energy, this can't be sustained for an unlimited amount of time (E. Anne Bardoel and Robert Drago, "Acceptance and Strategic Resilience: An Application of Conservation of Resources Theory," *Group & Organization Management* 46, no. 4 [2021]: 657–91).

16. Thank you to the beta reader who offered this analogy!

But the reality is quite literally the opposite. Some experiences that have been shown to reduce our capacity to cope include discrimination, mental illness, chronic illness and disability, and my personal favorite, being female (due to the challenges women face in the world). Moreover, prizing endurance without considering the unique hardships experienced by minorities—youth in foster care, people of Indigenous heritage, LGBTQIA+, soldiers and their families, and people with disabilities or chronic illnesses—can silence these groups from speaking up about their experiences, which can have dire consequences.

For example, I once became horribly sick on a family vacation but powered through because I was trying to be a good sport. Two days after returning home, I was admitted to the hospital with life-threatening complications, only to regret my misguided goal of powering through. And going back to Nietzsche, here's another fun fact: he experienced chronic, excruciating pain throughout his life. "It is such a strain," he once wrote, "getting through the day that, by evening there is no pleasure left in life. . . . It does not seem worth it." Really, it was only a matter of time until the poor man broke!

To summarize, some people have more resilience than others, but we all have our limit. So while we can and should boost our capacity as best we can, we must also remember that, like the budworm-ravaged forests of the late 1970s, every system has a breaking point after which it is exceedingly difficult to bounce back. More broadly, we'd all do well to remember that there isn't one "right way" to respond to adversity—and falling short on resilience is rarely the personal failure we perceive it to be.

CHAPTER 2: KEY TAKEAWAYS AND TOOLS

"Resilience did not, in fact, predict whether people became better and stronger after crisis."

- **Resilience**: The capacity to cope with hard things. A powerful tool to keep us together during shorter-term crises, rather than a singular strategy for coping with challenges long term.
- **From resilience myth to resilience reality**:
  1. Resilience isn't intended to help us thrive, but to help us survive.
  2. We can't always dramatically improve, or even control, our level of resilience. Some people have more than others, but everyone has their limits.
  3. Ongoing or extreme stress usually doesn't make us stronger; instead, it depletes our resilience and make us more vulnerable to breaking.
- **Grit gaslighting**: A phenomenon where we, or others, question our coping skills when we are cracking under the weight of our stress.

Chapter 3

Hitting Our Resilience Ceiling

> "Although it is entirely appropriate for us to grasp hold
> of the optimistic promise of . . . resilience . . . [it is not the]
> single answer to life's problems."
> —Resilience researcher Michael Rutter

The invitation announced an unfamiliar location: Naushon, an island off the coast of Massachusetts. Curious, I googled it. Apparently, my cousin Ryan and his fiancée, Kaitlin, were having their wedding on Kaitlin's family's *private freaking island.*

What?!

By any objective measure, I was in no shape to attend a fabulous destination wedding, let alone one on a fabulous but remote island nearly two thousand miles away from my home in Denver. In recent months, to manage my worsening pain, I had taken refuge in a special, adjustable zero-gravity bed, dubbed "The Eurich Group's Global Bedquarters." When I'd purchased the bed, my closest friends roundly mocked me for it ("Are you a seventy-five-year-old retiree?!"). But each morning, I'd wake up, hit the zero-G button—*whrrrrrrrrrrr!*—and start my workday thinking, *Who's laughing now?*

Though nearly all of my waking hours were spent in my zero-G sanctuary, pain still limited my workdays to five or six hours. Beyond the pain, I was pushing through impenetrable brain fog, often staring at my computer in confusion and overwhelmed by even the simplest tasks.

And yet, I still had three jobs (consulting, speaking, writing this book), which couldn't be done in six hours a day. Sleep soon became a casualty, sacrificed on the altar of my anxieties about staying afloat.

Then I started fainting. A lot. The first time it happened, I'd gotten up in the night to use the bathroom, and was found unconscious and sprawled on the floor. That's when I started using a cane to get around.

But even if it killed me, there was no way I was missing Ryan and Kaitlin's wedding. I imagined inhaling the salty air amid breathtaking ocean views and luxuriating in huge, fluffy beds—maybe there was even a spa!—while scores of staff milled about, tending to guests' every need. *Exactly* what my weary body and soul needed.

Getting there was a bit of a production: a flight to Boston, an overnight stay, a four-hour Peter Pan bus, and a ferry ride. As I met up with various family members along the way, I did my best to put on a brave face while cursing myself for leaving my cane at home.

When the ferry docked at Naushon, relief washed over me. I spotted a pickup truck, its engine idling, awaiting our arrival. Beyond the dock were several gorgeous houses along a wide unpaved road—or, more accurately after four days of rain, a vast mud pit. We were greeted by Kaitlin's lovely family, who told us they'd drive our luggage to our house, which "wasn't more than a fifteen-minute walk." (That truck? Yeah, it wasn't for us; it was for our luggage.) My stomach dropped. I hadn't walked fifteen minutes in months, let alone waded through a muddy quagmire. I grabbed my younger sister's hand as we wistfully watched our luggage drive away in style.

This was my first clue that this experience would be a bit more rustic—and *spaless*—than I'd expected. And the rain did not help one bit. We spent the weekend trudging miles through the mud to various far-flung locations across the island. Despite thrice-daily DEET baths, I was instantly covered in bug bites. (While bug bites are a mere annoyance for most people, for me, they mean enormous, painful welts requiring two times the recommended dosage of antihistamines.) Through my pain and exhaustion, the act of talking to so many people while trying to be pleasant and charming took its toll. At one point, when a well-meaning relative wouldn't stop interrogating me about my mysterious symptoms, I raced to the restroom to hide my frustrated tears.

Miraculously, the rain let up briefly during the ceremony, which was beautiful. But like clockwork, just as we started trekking from the wedding venue to the reception—"a quick mile"—the sky opened up, *biblically*. Upon reaching the reception tent, someone offered to escort me back to the house. Teeth rattling, I refused. If I had to walk even one more step, I knew I would collapse right then and there.

The trip home involved lots more rain and a canceled bus. But I resiliently forced my body—exhausted, hive-covered, stinging with pain—to keep moving. Upon returning home, I celebrated by sleeping for three days. On the fourth day, I woke up for a scheduled call with a colleague. And finally, from the cocoon of my zero-G bed, I found the courage to admit—out loud to someone else—that I had reached my breaking point.

As long as I live, I will never forget his response.

"Wow, Tasha. I know you're going through a lot. But I never imagined you'd be taking things *this hard*. That's surprising to me."

Normally, a comment like this would have sent me into a spiral of shame and self-flagellation for not being "tough enough." Instead, what happened was that I gave up. Right there in zero-G in my sushi pajamas. "Screw it," I said to no one, "I can't do this anymore."

While I didn't know it at the time, that was the moment my resilience officially ran out. The following months were a blur. I stopped checking the boxes in my resilience spreadsheet and then printing them altogether. I spent my truncated workdays and quick keynote trips pretending I was fine. Then, I'd close the curtains, pull up the covers, turn on the TV, and eat and drink myself into oblivion.

I knew this was making everything worse. But truthfully? I no longer had it in me to care.

WHEN DEMAND EXCEEDS SUPPLY

As remarkable as the human capacity for resilience can be, as we learned in Chapter 2, it is not an inexhaustible resource. Per resilience researcher

Michael Rutter, it's "biologically implausible" for resilience to steel us against all stress. As evidence, in a survey of my newsletter subscribers,[1] more than 75 percent believed that while they'd managed to adapt to their stress so far, it was becoming harder over time.

Imagine for a moment that you're back in grade school, listening to a teacher drone on about a subject you neither care about nor understand. Simultaneously bored and lost, you grab a rubber band from your bag and absentmindedly start stretching it like a makeshift fidget spinner. That whole semester, it becomes your constant class companion and quiet stress reliever. Then, one day, the unexpected occurs: the rubber band snaps. Months of stretching and rebounding didn't make it stronger; they wore it down over time.

Our resilience is a lot like that rubber band—a decent way to manage stress for a while, maybe even a longish while. But eventually, after bending and stretching week after week and month after month, something in us snaps. These moments can be grand, dramatic meltdowns or quiet flashes of despair, but they almost always come without warning and shatter us in an instant. I call this phenomenon, where we reach the upper limit of what we can endure, **hitting our resilience ceiling.** This is our breaking point, where our capacity to cope has become so depleted that we snap at the slightest setback, demand, or annoyance. To repeat a point from Chapter 2, some people have higher resilience ceilings, and others have lower ones, but *everyone has a limit*. And, as I learned, we often don't know we even *have* a resilience ceiling until we've hit it and are feeling its effects, like poorer mental and physical well-being, less successful coping, and more distress, discouragement, and shame.

The same is true for grit, a close cousin of resilience, defined as the passion and persistence to pursue long-term goals. As grit pioneer Angela Duckworth and her colleagues have expertly documented, grit can lead to myriad positive outcomes. But like resilience, this capacity is not unlimited. If we rely on grit too extensively, we risk investing too much

---

1. I hope you'll join our wonderful community, where you'll receive monthly insights to help you be the best of who you are and what you do (and no spam, ever). Just go to www.tashaeurich.com to sign up!

energy and effort toward goals when changing direction, or even quitting, is the better path for our well-being.

For anyone who prides themselves on their mental toughness, hitting one's resilience ceiling can be especially disconcerting. If you're anything like me, it feels shocking, even shameful, to be so suddenly beaten down by adversity—especially when you're trying so hard *not* to be. When I asked a few hundred high achievers in a client organization (a Big Four accounting firm) what hitting their resilience ceiling feels like, many mentioned frustration, anger, impatience, or annoyance. Others felt unmotivated, hopeless, numb, frozen, or disconnected (as in, "I just *can't*"). Some reported crying uncontrollably or exploding at the slightest annoyance.

This phenomenon might sound a lot like **burnout**, the emotional exhaustion, detachment from others, and lack of accomplishment that esteemed psychologist Christina Maslach defines as a reaction to excessive work stress, and researcher, leadership expert, and my dear friend Liz Wiseman attributes to too much work with too little impact. But while burnout can deplete our resilience resources and push us closer to our ceiling, the two experiences differ in onset and scope. First, burnout develops gradually, wearing us down bit by bit, while hitting our resilience ceiling feels sudden, like we're breaking on impact. As one client aptly described it, "you're fine . . . until the second you're not."

Further, burnout is specific to work, where our resilience ceiling is a product of the total stress we're experiencing across all areas of our lives—we can therefore hit our resilience ceiling even when we're not burned out, typically because of compounding stressors across multiple life domains (work, family, friendships, etc.). Alternatively, we might burn out at work without hitting our resilience ceiling in our lives as a whole.

The more we tap our resilience reserves without taking the necessary time and space to replenish them—usually through good old-fashioned "doing as little as possible"—the more vulnerable we become to the stressors that depleted our resilience in the first place. And in today's chaotic world, you might be closer to your resilience ceiling than you realize. Research shows that we tend to overestimate our coping capacity; that is,

we tend not to be self-aware about our resilience. For instance, one survey revealed that 83 percent of Americans saw themselves as highly resilient, but only 57 percent actually scored high on validated measures of resilience. This overconfidence is especially risky in times of stress, when our resilience reserves are being tested. In this way, overrelying on resilience can become a source of fragility.

There is an excellent metaphor from the chronic illness and disability community, known as **spoon theory,** that illustrates why this is the case. Christine Miserandino (who lives with lupus) first came up with this concept when out to dinner with a friend. When her friend asked what it was like to live with her disease, Christine grabbed about a dozen spoons off several nearby tables to explain. Healthy people, she began, especially young healthy people, start their days with unlimited spoons. But people with disabilities start the day with only a handful, forcing us to strategically manage our energy. "The difference between being sick and being healthy," she explained, "is having to make choices that the rest of the world doesn't have to." She then handed her friend twelve spoons, announcing, "Here you go. You have lupus."

Christine asked her friend to rattle off the tasks she needed to complete on a typical day after waking up, even simple ones. For each—showering, getting dressed, making breakfast—Christine grabbed a spoon back. Her friend was surprised that, by the time she arrived at work, she was already down to six spoons. "That's why you have to choose the rest of the day wisely," Christine explained. "You never know when you'll need a spoon. And when your spoons are gone, they're gone."

I've seen firsthand how chronic illness means managing trade-offs and saying no to an annoying number of things. In discovering how to navigate my new life, slowly but surely, I've learned to ask, "Do I *really* want to spend a spoon on this?" It's the same with resilience. Some of us might start with more resilience spoons than others, but we all have a limited number. And with each spoon spent, we inch closer to our resilience ceiling—so we'd *all* do well to use them wisely.

Because while a singular reliance on resilience may have served us in more stable times, humans simply weren't designed to endure the kind of endless disruption we experience today. It is therefore possible that, both individually and collectively, we're hitting the upper limit of what our "grit and bear it" paradigm can do for us. Alarmingly, almost half of my Big Four accounting firm employees reported that their primary coping tools were losing effectiveness, and only 14 percent reported being nowhere near their resilience ceiling.

Anytime you notice that your best coping tools aren't cutting it, that's a flashing warning sign that your resilience is running out. When this happens, the solution is *not* to push yourself harder—it's to do less. The closer you are to your resilience ceiling, the more you'll need to release your pressure valve. Your goal should be to offload whatever you can and decompress as much as possible.

So, how close are you to your resilience ceiling? For a more comprehensive answer, check out the free Resilience Ceiling Quiz at www.Resilience-Quiz.com (and feel free to share it with your friends!). For now, here are a few simple clues:

Clues	How It Feels	Real-Life Examples
1. Lost mojo	• Having less energy and motivation to keep all the plates spinning.	• "Balancing everything is more exhausting than I can remember."
2. Little things feel big	• Getting unusually worked up about relatively minor issues.	• "Why did I just blow up at my spouse about how they load the dishwasher?"
3. Top tools failing	• Feeling like your go-to coping strategies are one more thing to do instead of a source of relief.	• "I've been trying to stay hopeful and optimistic, but life is still wearing me down."

WHEN RESILIENCE ISN'T ENOUGH

The day Baratunde Thurston's mother went out to run some errands, the eleven-year-old was having an ordinary day, riding his bike around his neighborhood with his friends.

Then, he spotted Michelle, the object of his adolescent affection, behind the wheel of a passing car. Fueled by prepubescent excitement, young Baratunde pedaled as hard as he could in an effort to catch up with the car. Maybe then, his best friend's older sister might finally notice him.

But fate had other plans. A malfunction in his bike's gears sent Baratunde tumbling over his handlebars and onto the pavement. He felt a searing pain on his face and looked down to see blood gushing onto the road.

Luckily, Baratunde was only a few blocks away from home, and he had his keys, so he returned to his empty house and got to work tending to his injury. Forgoing proper wound cleaning, he reached for some kids' play putty, flattened it with his palm, and pressed it over his wound to stop the bleeding. Then, gingerly placing an ice pack on his throbbing face, he wrapped himself in a cocoon of blankets to stave off the chills. All he had to do now was patiently wait for his mother to appear in the doorway.

When she finally did, Baratunde recounted his harrowing incident.

"Are you *okay*?" she asked, concern etched across her face.

"I'm good. I've got it under control," he confidently replied. Through his worsening pain, the resourceful eleven-year-old felt a surge of pride—an overwhelming feeling of accomplishment for having toughed it out and tended to his injury without anyone's help.

In that moment, Baratunde actually needed his mother more than he could admit—not only to clean his wound and affix a proper bandage but to offer the comfort an eleven-year-old needs after a nasty fall off his bike. And yet, he put on a brave face. From a young age, the future *The Daily Show* correspondent had been conditioned to believe that soldiering through life's mishaps was a badge of honor. As a Black child of a single mother, Baratunde was especially vulnerable to societal expectations to be strong and self-reliant, leading him to develop **skin-deep resilience,** a phenomenon where people display outward strength despite internally breaking. Unsurprisingly, this worsens stress and carries substantial physical and psychological consequences.

It's easy to see how the pressure to persevere depletes our cognitive resources—paradoxically, this makes it even more difficult to sustain our resilience over time. It also explains why resilience practices don't always recharge us. Take, for instance, my resilience spreadsheet: looking back, the pressure I put on myself to check *every* box *every* day added to my stress instead of alleviating it. And on days I fell off the bandwagon, I felt guilty, which further drained my mental energy. This pressure to engage in practices like self-care goes a long way in answering the question posed in the headline of a recent *Atlantic* article: "How Did Healing Ourselves Get So Exhausting?"

It wasn't until many years later that Baratunde came to understand that **his singular focus on resilience came at the cost of honoring and advocating for his own needs.** Sure, he'd learned to push through pain, but in so doing, he'd neglected many vulnerable parts of himself that were yearning to be seen. This habit of repressive coping is a common co-traveler to skin-deep resilience. While avoiding hurtful thoughts, emotions, and experiences may protect us from psychological pain in the short term, it costs us dearly in the long term (more on that in Chapter 5). In Emmy Werner's children of Kauai study, for example, although most resilient children managed to build relatively successful and functional lives compared to their less resilient counterparts, they often showed a persistent need to detach themselves from their feelings—translating to a fear of long-term commitment, "a certain aloofness in their interpersonal relationships," unhealthy overachieving, and stress-related problems like migraines and backaches.

When we steadfastly forge through our challenges, unwavering in our resolve to overcome them at all costs, we are often praised for our grit and persistence. But speaking from personal experience, this **costly persistence** can lead us to deny negative emotions, downplay harsh realities, and tolerate intolerable situations—all of which rob us of agency and diminish our motivation to change the things that we can. Or as Baratunde powerfully observed, "When I [move] so quickly to resilience, I miss a chance to fully feel into my needs."

Back when I hit my own resilience ceiling, it became abundantly clear that coping and carrying on weren't working anymore. If only I could supplement my resilience tools with a second, complementary approach—something that would take me beyond merely managing my stress and suffering and allow me to truly tackle their underlying causes. But what? I was sure the answer was somewhere in my data, specifically in that group of people who learned how to thrive in adversity despite being no more resilient than anyone else.

Indeed, the answers *were* there—but they wouldn't truly crystallize until I met an extraordinary leader named Nabeela Elsayed. Nabeela showed me that while what doesn't kill us doesn't automatically make us stronger, it *can* wake us up to a better way.

CHAPTER 3: KEY TAKEAWAYS AND TOOLS

"You're fine . . . until the second you're not."

- **Key Takeaways**
  - Resilience is an exhaustible resource.
  - Overreliance on resilience can be a source of fragility.
  - Defaulting to resilience can mean missing the chance to fully feel into our needs.
- **Hitting our resilience ceiling**: When we reach the limit of what we can resiliently endure, the slightest setback, demand, or annoyance can break us. Clues you're close to yours are as follows (and don't forget to check out the free Resilience Ceiling Quiz at www.Resilience-Quiz.com):
  - **Lost mojo**: You have less energy and motivation to keep all the plates spinning.
  - **Little things feel big**: You're getting unusually worked up over minor issues.
  - **Top tools failing**: Your best coping strategies feel like piling on instead of providing relief.
- **Resilience spoons**: A metaphor to imagine the limits of our resilience. We all have a limited number and should spend our spoons wisely.
- **Skin-deep resilience**: Showing outward strength when we're inwardly breaking.
- **Costly persistence**: Continuing to push through challenges despite what it's costing us.
- **Resilience Dos and Don'ts**

Dos	Don'ts
• Draw on resilience as a short-term tool for coping with sudden, out-of-control situations.	• Make resilience your singular, long-term coping strategy.
• Conserve your resilience resources.	• Resiliently push through when you're close to your resilience ceiling.

Chapter 4

The Second Skill Set

> "In a chronically leaking boat, energy devoted to changing vessels is more productive than energy devoted to patching leaks."
> —Warren Buffett

Nabeela Elsayed isn't just a high achiever; she is a veritable rock star. With a razor-sharp business mind, she wields dual degrees in marketing and supply chain management and distribution and a master's in industrial/organizational psychology. Beyond an impressive string of accomplishments in her stellar retail career, the young professional has served on the boards of TaskRabbit and the Miracle Foundation (a charity for orphans) and held advisory roles at the Future Skills Centre of Canada.

In her, Nabeela's colleagues see someone who's always on top of things no matter what fresh chaos is exploding around her. In the toughest situations, she is everyone's go-to—always ready to confront any challenge, like a Marvel superhero in business casual.

But there's something else that Nabeela is. She's human. Since the earliest days of her career, like most people, she'd suffered from her fair share of anxiety and self-doubt. She remembers working as a store manager when she first felt the sensations she'd quickly learn to dread: the racing heart, the shallow breathing, the hot prickles in the skin of her wrists and up the back of her neck. The first time it happened, she could hardly believe what was happening. It felt like being throttled by a ghost.

Nabeela was having an anxiety attack.

Instead of asking for the help she needed, she chose to conceal her struggles. Even though she was best friends with her boss and working for a company she adored, reaching out would have revealed her vulnerability, and that wasn't an option. Instead, she decided to safeguard her hard-earned reputation by gritting and bearing whatever came at her.

Nabeela eventually ascended to the role of global head of digital and retail transformation at Swedish retailer Ikea, where she was tasked with turning its brick-and-mortar stores from retail locations with regular hours and customer-facing staff to twenty-four-hour e-commerce fulfillment centers. The overhaul demanded a complete reimagining of their systems, processes, and workforce. Meanwhile, the role had meant moving her family to Sweden, which introduced a fresh set of challenges across Nabeela's whole life. "My stability at home kind of fell apart," she explains. "Suddenly, there was no respite, and my marriage started suffering."

Before she knew it, Nabeela found herself steering a global transformation for one of the world's most iconic brands alongside the initial stages of a divorce. It was during this time that she found herself suffering an unprecedented wave of anxiety attacks. In a mode of constant coping, she desperately tried to hold everything together and forced herself not to crack. On the bright side, no one at work had any idea how much she was suffering. "Everyone was amazed at what I was doing," she says. "I was concealing my pain like a *boss*."

During this ordeal, Nabeela was invited to a company retreat with Ikea's top 250 leaders in Malaga, Spain, where she was asked to update the senior team on the extensive changes she'd been spearheading. Arriving at her hotel after a three-hour train ride, she felt like she was balancing on a tightrope in a windstorm. She just had to keep smiling, deliver her presentation, and get back home to her kids before everything came crashing down.

Then came the meeting. Nabeela sat there, surrounded by a sea of faces that were clearly worried, confused, even outright furious. Suddenly, a barrage of complaints ensued. Nabeela's colleagues told her that not only was her project failing but also that she was damaging the very company she loved and to which she was giving her all.

That night, Nabeela shut herself in her hotel room and tried not to fall apart. She had sacrificed every last part of herself—her happiness, her sanity, maybe even her marriage—for this project. And her teammates were not only unappreciative of what she was doing; they were actually angry with her! Tomorrow would be another day of meetings, and she simply couldn't go on like this. Something *had* to give.

The following morning, Nabeela stood nervously in front of her colleagues and cleared her throat. In a fragile voice, she told them that as a twelve-year associate, the last thing she wanted was to take their beloved company down a path that risked its culture and values. Then, Nabeela did something that she had never done before. Not once in her entire career.

She set aside her hard-won image of invincibility and courageously confessed to the deep personal pain she felt in hearing her colleagues' criticisms. "It hurt," she admitted. "Because I'm not the rock you think I am. I've been struggling. Believe it or not, I've had to seek therapy to get myself through this."

This was Day One of Nabeela's transformation. In the months ahead, she would go on to shed the superhero persona she'd spent her entire life carefully crafting. "I realized I had a really powerful limiting belief," she told me, "that I could do anything and always handle more. Intellectually, I knew that this belief was limiting me. I had to break through it to move myself forward." Armed with this key insight, she started to summon the courage to change.

From there, Nabeela embarked on a journey of self-renewal. She started exploring new ways of working and actively road-testing them. She reached out to the colleagues who'd criticized her and asked them how she could do better. This vulnerability fueled a newfound candor: with her colleagues now feeling safe to voice their concerns, conversations were suddenly freer, which helped Nabeela refine and implement her project. "Anytime you widen the sounding board," she observes, "you're gonna make better choices." The results speak for themselves. The company exceeded sales growth and employee engagement targets, nearly tripled its e-commerce, and was recognized by the likes of *Harvard Business*

Review, *HR Digest*, and McKinsey for the incredible success of its digital transformation.

But while Nabeela was proud of beginning to master the habit of asking for help, she found herself unable to *completely* shake her "always on top of things" image—and it was coming at too high of a cost for her mental and emotional well-being. **She knew she had to find a way to start truly honoring, and advocating for, her own needs.**

After more reflection, Nabeela decided it was time for a change. She bid farewell to Ikea and relocated to Toronto, taking on a senior role at Walmart Canada, overseeing ninety thousand associates. Somewhere over the Atlantic, while she was drafting a plan for her first one hundred days as Chief People Officer, she kept scribbling the same two words: "Stop pretending, stop pretending, stop pretending." Then it hit her. This new job was her chance to finally let go of who she was before.

This was the opportunity she'd been waiting for.

Days later, she addressed a group of several thousand new colleagues. Introducing herself as a recovering workaholic, Nabeela shared the steps she'd taken to transform: recognizing the costs of covering her feelings, learning to stop being afraid of what other people thought, and radically reprioritizing her life to support her long-term success and well-being.

Today, Nabeela has no doubt that breaking free from these self-limiting goals and habits, and building new ones in their place, was what helped her finally step into her own power. "Authenticity is the strength I've grown the most," she notes. "I can speak genuinely and honestly about myself, and it resonates with people."

After that, with a string of major life events in quick succession, Nabeela's new habits were put to the test when she remarried, had her third child at age forty, and was promoted to Walmart Canada's chief operating officer (COO) as only the third woman in the company's history. But while "resilient Nabeela" might have buckled under the stress, "new Nabeela" was energized. No longer wasting energy covering up her emotions, her former stress and overwhelmed feelings were replaced with a welcome sense of peace. "I guess the word that comes to mind is just relief," she explains. And the anxiety attacks? They're now a thing of the past.

It would be logical to conclude that Nabeela bounced back from her personal and professional problems by tapping into a new reservoir of resilience. But upon hearing her story, I instantly knew that what she was describing wasn't resilience—in fact, her commitments to mental fortitude and invulnerability were what had caused so many of her problems in the first place!

That's when it hit me: "new Nabeela" displayed all the characteristics I'd identified in that mysterious third group of interviewees who thrived through tough times. For the first time in a long while, I saw a light at the end of the tunnel—not just for my research, but for my life too. Feeling more motivated than I had in months, I *again* revisited all the data I'd collected, determined to figure out what set that third group of participants apart.

Soon, I finally had my answer.

THE BREAKTHROUGH BLUEPRINT

A few months ago, a friendly repairman visited my loft to adjust a newly installed glass shower door. I led him to the bathroom and left him to his work. A while later, a series of loud bangs reverberated down the hall.

Alarmed, I rushed back to check in. Noticing the concern on my face, the repairman smiled reassuringly, "Don't worry. Glass can withstand a surprising amount of abuse. You need to hit it just right, or repeatedly over time, for it to crack."

This insight rang true to me. Several of the wine glasses in my house have been dropped (often by me) from several feet without sustaining a single crack, but when I've dropped my iPhone from a couple of inches, it has shattered completely.

Likewise, I've always been amazed at both how strong *and* fragile the human spirit can be. Sometimes, like the wine glass dropped from several feet, we can withstand tremendous stress, pressure, and adversity without breaking. Other times, the tiniest setback or stressor bearing down can shatter us like an iPhone screen. How can both of these things be true? Probably like you, I've dropped many a phone. Have you noticed that they rarely break the first, second, third, or even *tenth* time you drop the darned thing? But even as we marvel that they have survived another nasty fall in one piece, weak spots are still forming underneath—invisible to the naked eye and just waiting to shatter with the slightest touch and in the most unexpected moment. Much like glass, with each setback or crisis we resolutely endure, our resilience weakens, often imperceptibly, with each event leaving us more vulnerable to shattering.

What if there was a better way? What if we could learn to spot these weak spots earlier and take steps to repair them long before we're on the brink of breaking? Or even when we *are* breaking, what if we could repair these cracks in a way that makes us even stronger than before? This is the concept underpinning **kintsugi,** the centuries-old Japanese art of mending broken objects using a lacquer infused with a precious metal like gold. Kintsugi is believed to have originated in the late fifteenth century under Shōgun Ashikaga Yoshimasa. As the story goes, when one of his ancient Chinese bowls broke, he sent it back for repairs, whereupon it returned reassembled with ugly metal staples.

Unimpressed, the feudal chief challenged his craftsmen to find a better solution. Why not use gilded lacquer? It would highlight the cracks instead of concealing them, while the intricate gold veins would render the piece even stronger. Hundreds of years later, in Japanese culture, these "cracks" are seen as an integral part of life—and embracing them is believed to enhance their beauty.[1] What if our own cracks were actually features rather than flaws?

One simple yet profound insight revealed by Nabeela and our third group of interviewees is that the best response to hard things isn't merely to *survive* them; it's to *harness* them. People who are strengthened by adversity don't try to conceal their cracks with a "powering through" mindset or by denying their broken parts. Instead, they harness them to become more beautiful than before. **Becoming shatterproof, at a fundamental level, means harnessing the broken parts of ourselves to**

---

1. I recommend entrepreneur Audrey Harris's fascinating 2015 TEDx talk on this topic.

access the best version of ourselves. After all, as both Rumi and Leonard Cohen remind us, "The crack is where the light enters."

On April 12, 1959, then-senator John F. Kennedy addressed the United Negro College Fund convocation in Indianapolis, Indiana. He painted a picture of a world in crisis, describing the "most urgent peril history has ever known." Then Kennedy offered an eloquent metaphor: "When written in Chinese, the word 'crisis' [has] two characters—one represents danger, and one represents opportunity." Inspiring words, surely. There was just one problem. His translation wasn't entirely correct.

The real translation, I think, is far more interesting. According to Chinese language and literature professor Victor Mair, the Chinese word for crisis (*wēijī*) is made up of two syllables, each with their own character: *wēi* (危) and *jī* (機/机). While *wēi* does signify a dangerous moment, *jī* doesn't exactly translate to "opportunity"; it's more like a turning point when something "begins or changes" and when, depending on our actions and choices, things can turn out for the better *or* the worse.

Indeed, crisis can foster opportunity, but not automatically. Especially if we're already starting to strain under the stress of prolonged uncertainty or chaos, even the smallest challenge can threaten to break us. However, if we harness adversity, interrogating and channeling it for personal reinvention, we'll discover new opportunities in even the darkest times. On this point, Buzz Holling would likely agree. He calls crisis "both creative and destructive," in that it "destroys accumulated rigidities" *and* fosters "new opportunity." But, he stressed, we can only realize the opportunities crises present by embracing transformation—starting with resisting our human tendency to protect the status quo.

It's easy to see how this could be true for many common challenges, like Nabeela's career setback or her divorce, or young Baratunde's bike crash. But what happens when our crisis comes in the form of an acutely traumatic or life-threatening event—like betrayal, abuse, or violence—something so stressful that it risks leaving us forever traumatized and vulnerable? Here too, there is hope. In the late 1990s, psychologists Richard Tedeschi, Crystal Park, and Lawrence Calhoun found that for some people, even the most traumatic experiences fostered growth. Through interviews and surveys with trauma survivors, they identified several ways individuals can emerge with a new appreciation for life, renewed strength, deeper relationships, and spiritual development. The researchers termed this phenomenon post-traumatic growth.

Which brings us to the big idea: **to flourish in a world of constant chaos, we must go beyond surviving or even adapting and learn to harness adversity as an opportunity to transform.** As we'll soon see, this journey doesn't require us to master any particular skill. Instead, by following the simple four-step process you'll learn in Part Two, almost anyone can build a more shatterproof life.

BEYOND RESILIENCE

On the surface, becoming shatterproof might still sound a lot like becoming more resilient. But the two concepts are fundamentally different in four important ways. First and foremost, a resilience mindset assumes that what doesn't kill us automatically makes us stronger, so long as we simply survive. A shatterproof mindset, on the other hand, maintains that what doesn't kill us will make us stronger *if and only if* we choose to harness it as a force for change. In other words, resilience is reactive while becoming shatterproof is not just active—it's *pro*active.

Second, resilience is largely a defensive strategy, focusing more on endurance and recovery than upping our game. So, while resilience can get us back to our baseline levels of happiness and well-being, becoming shatterproof gives us access to our best selves—with tangible improvements in meaning, personal growth, inner peace, and even physical health.[2]

Third, traditional approaches to resilience typically offer a one-size-fits-all approach[3] that attacks all varieties of stress and unhappiness with

---

2. These findings are consistent across culture (Shi Yu, Chantal Levesque-Bristol, and Yukiko Maeda, "General Need for Autonomy and Subjective Well-Being: A Meta-Analysis of Studies in the US and East Asia," *Journal of Happiness Studies* 19 (2018): 1863–82).

3. One example of a newer and more targeted approach to building resilience can be found in George Bonanno's book *The End of Trauma*.

the same practices like self-care and optimism; in contrast, shatterproof practices are targeted to reduce frustration of, and better fulfill, specific unmet psychological needs. In support of that distinction, recent research suggests that resilience isn't a "primary nutrient of growth and wellness," and maximizing well-being requires going beyond resilience to specifically enhance need fulfillment.[4]

Finally, while resilience drains mental resources, shatterproof practices replenish them. Instead of the exhaustion that comes from drawing down our limited reserves, we'll feel more energized, enthusiastic, and motivated. Conveniently, this means we can turn to the Shatterproof Road Map anytime—especially when we feel one setback away from breaking—to help us work through our stress in new and helpful ways.

For anyone who's always prided themselves on their grit and resilience, this might feel like a radical shift in thinking, and indeed it is. Therefore, to kick-start our shatterproof journey, we should lay the groundwork with three important mind shifts.

THREE SHATTERPROOF MIND SHIFTS

Mind Shift 1: From Discounting to Embracing Pain

The first mind shift is moving from minimizing hard things to actively embracing them. Think back to Nabeela's story: until her pivotal moment in that Spanish hotel room, she saw the admission of emotional pain as a personal failure. Her transformation became possible only when she began paying attention to these emotions—to see how close she was to breaking and how much her invincible persona was hurting her. By acknowledging her true feelings rather than pretending they didn't exist, Nabeela took the first step toward personal reinvention

---

4. There's some evidence that frustrated psychological needs might be a more significant influence on our mental health than trauma. In one study with Darfuri refugees, their distress levels were more strongly related to need frustration than the war trauma they'd experienced (Andrew Rasmussen et al., "Rates and Impact of Trauma and Current Stressors among Darfuri Refugees in Eastern Chad," *American Journal of Orthopsychiatry* 80, no. 2 (2010): 227–36).

(and so can we).[5] In this sense, **becoming shatterproof is self-awareness walking.**[6]

Mindshift 2: From Coping to Courage to Change

The second shift requires us to move our focus from coping to finding new courage to reinvent ourselves. For years, Nabeela followed the resilience playbook, diligently proving that she could handle any challenge in her path. Yet this "just stay strong until things improve" mindset was a temporary Band-Aid that didn't address the root causes of her stress. By relying solely on resilience, we not only overlook the deeper issues causing us preventable pain[7] but also risk missing vital, actionable insights that can motivate us to shake things up for the better.

When we've spent a long time doing anything—in this case, resiliently coping—inertia often tethers us to the old way. As mega-bestselling author and the world's top leadership thinker Marshall Goldsmith (who also happens to be my friend, mentor, and honorary dad) has said, "The most reliable predictor of what you'll be doing five minutes from now is what you're doing right now." This is because inertia is easy and change is hard. In the same way you can't bear to part with that twenty-year-old sweatshirt from high school that's covered in holes (the one you love even though it doesn't keep you warm anymore), it probably feels easier to stick with comfortable behaviors even though they've stopped helping you.

Yet especially when we've run out of the capacity to cope, we must find the courage to change. Where resilient people stay the course, shatterproof people grow—and ultimately discover that **change is pain repurposed.**

---

5. Confronting our imperfections is equally important to teams and organizations as it is for us as individuals. A three-decade McKinsey study of over one thousand companies across fifteen industries uncovered one factor that differentiated those that maintained their competitiveness over extended periods. Instead of reactively adapting to market conditions, they proactively transformed by regularly questioning their current state, reimagining their future, and reconfiguring (Richard Foster and Sarah Kaplan, *Creative Destruction: Why Companies That Are Built to Last Underperform the Market—And How to Success Fully Transform Them* [New York: Currency, 2011]).

6. A smart beta reader said that this reminded them of my colleague Susan David's gem: "Courage isn't the absence of fear; courage is fear walking."

7. Is all pain preventable? NO!

Mind Shift 3: From Bouncing Back to Growing Forward

The final mind shift is replacing the goal of retaining or regaining what we had with the goal of becoming better, stronger, and mentally healthier than before. After all, why are we fighting so hard to merely bounce back when we were *barely* getting by? Rather than reverting to old, familiar habits that weren't serving us, what if we could review and renew ourselves by choosing new goals and habits to propel us forward? Consider Nabeela's **grow forward plan**. It was based on a new mindset that her needs mattered *and* she would find new ways to fulfill them:

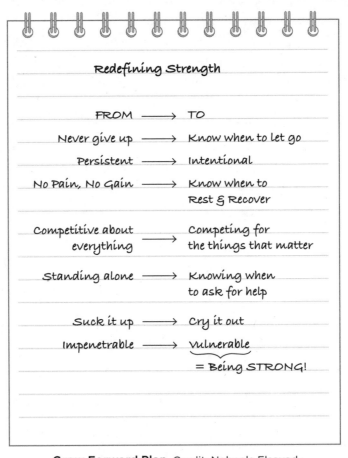

Grow Forward Plan Credit: Nabeela Elsayed

As another dear mentor and friend, Alan Mulally,[8] says, even when we don't have everything figured out yet, we must always believe a better way exists—something he calls the "**better way mindset.**" This choice requires bravery, but the payoffs are huge.

STARTING YOUR SHATTERPROOF JOURNEY

By this point, it should be abundantly clear that Nabeela's journey wasn't mere evolution or incremental adaptation—it was pure, proactive transformation. Using her stress and frustration as data, she was able to identify the habits holding her back, which helped her carve a new path of growth, well-being, and purpose. This is precisely what happens when we move beyond resilience to become shatterproof.

Before we dive into the four steps to get there—also known as the **Shatterproof Road Map**—let me address one reasonable objection. If you're a busy, in-demand person, you might be thinking, "Tasha! I don't have time for transformation; I can't even finish today's to-do list!"

I see you and I hear you.

Here's the thing: if you're reading this book, you're probably already investing some time and energy into coping strategies that you now know can have diminishing returns, especially when you're approaching your resilience ceiling. What if you took even *part* of that energy and reinvested it? Say you're a single parent who works full time, and you only have fifteen minutes to yourself in the mornings before your kids wake up. Maybe you've been spending that time meditating. If that's helping you, keep doing it! But if it isn't helping as much as it used to, or you're doing it out of "mustivation," what if you reinvested those fifteen minutes in reading more of this book, completing one of its assessments, or even more simply, trying to feel into your own needs in service of better well-being? As you'll see, becoming shatterproof doesn't have to be

---

8. Alan Mulally is one of the greatest CEOs of his generation. If you've read *Insight*, you know all about his incredible transformation of two iconic American companies. Either way, if you're a current or aspiring leader, do yourself a favor and read Bryce Hoffman's page-turner on Mulally's turnaround of Ford, *American Icon*.

time-consuming, and the more you prioritize it, the easier it gets. And anyway, as I advise every CEO I coach, there's nothing wrong with making personal development as easy as possible. In fact, it's a favor to your future self!

Even more fundamentally, the essential prerequisite to begin this journey is remembering the true cost of *not* changing. To that end, let me leave you with an ecological metaphor of two bird species. On one wing, we have the dodo, a once-thriving, flightless pigeon. For centuries, dodos lived a life of ease on the beautiful, remote island of Mauritius, a habitat abundant with food and scarce with predators—until the sudden arrival of humans shattered their peaceful existence. The dodos had never faced such a threat and never developed the ability to adapt. They became easy prey and were extinct in mere decades.

Contrast this with the tawny owl, a nimble bird native to regions that are experiencing abrupt changes in climate. Over many years, these opportunistic owls have developed the ability to adapt, having transformed their light-colored plumage into a dark-brown hue that seamlessly blends into less snowy surroundings. As a result, the tawny owl population isn't shrinking or even flattening; it's growing. The lesson in nature, and life, is clear: **in an ever-changing world, those who reinvent themselves don't just survive—they find new ways to prosper.**

Which would you rather be: the disappearing dodo or the adaptive owl?

CHAPTER 4: KEY TAKEAWAYS AND TOOLS

> "When we rely solely on resilience, we overlook the deeper issues causing us preventable pain, and risk missing vital, actionable insights that can motivate us to grow."

- **Becoming shatterproof**: Proactively channeling adversity to grow forward: harnessing the broken parts of ourselves to access the best version of ourselves.

Resilient	Shatterproof
Passive	Proactive
Restoring the status quo	Finding new ways to thrive
One-size-fits-all	Targeted to specific needs

- **The three shatterproof mind shifts**
    1. From ignoring hard things to proactively embracing them.
    2. From the capacity to cope to the courage to reinvent ourselves.
    3. From bouncing back to growing forward.
- **The Shatterproof Road Map**: Four simple steps to help you move beyond bouncing back and start growing forward (and the focus of the next section of this book).
- **Two tools to get you started**
    - **Better way mindset**: Choosing to believe that a better way exists, even if you haven't figured out what it is yet.
    - **Grow forward plan**: A one-page plan charting your proactive transformation using the prompt "I want to move from _____ to _____."

PART TWO

The Shatterproof Road Map

Chapter 5

Step 1: Probe Your Pain

> "All pain triggers a reminder, deeper than thought, buzzing through blood and bone, that we are fragile and finite."
> —K. J. Ramsey

In Chapter 2, we met Jan, the mother who endured grit gaslighting from her doctor and loved ones while navigating her son Andrew's behavioral issues. Now, let's fast-forward nearly a decade to one frigid winter evening when Jan was driving home in a snowstorm. As she pulled into the driveway, her relief instantly gave way to dread. There in her snow-covered backyard stood Jan's seven-year-old daughter, Ainsley, barefoot and shivering.

As she wrapped her coat around her daughter, she could feel Ainsley's tiny shoulders shaking with sobs. "Honey, why are you out here?" Jan asked. But she already knew the answer.

From the outside, the family's home might have seemed like a refuge from the heavy snowfall, but inside, another storm raged. Stepping through the front door, they confronted its epicenter: nine-year-old Andrew, in a full-on meltdown. His eyes blazing with fire, Jan's gentle, loving boy was unrecognizable as he screamed, ranted, and swore. Looking around, she spotted several holes he'd punched in their living room wall.

For Jan and her husband, David, these outbursts had become a horrifying new "normal." Andrew was never a calm child, but starting shortly after his ninth birthday, things worsened. Each day brought grueling, hours-long meltdowns and endless compulsive rituals, including an unexplained aversion to doorways and the compulsion to gnaw subway floors. Once, Andrew put a sharp knife in his mouth to, as he put it, "feel something." Days earlier, he'd made a heart-wrenching confession. "M-mommy," said the trembling boy, "I feel like I'm going crazy. I want to die." As he stood there looking up at her, Jan's heart dropped out of her body.

Meanwhile, his sister Ainsley had also developed behavioral issues that landed her in the principal's office almost daily. The week prior, she'd handed her mother a note scrawled in red crayon: "Mom, I am sorry I am a bad girl."

As her children's struggles intensified, Jan retreated further inside herself—her focus became surviving each minute. Acknowledging her anguish would surely break her, so she started suppressing her suffering. The strategy soon became effortless. She was so good at pushing down her fears and worries, it was almost like she felt nothing at all.

THE ART AND SCIENCE OF AVOIDANCE

Like Jan, most of us have experienced times when our negative emotions are so overwhelming that we disconnect from them entirely. Sometimes it's a conscious choice; other times, our circumstances leave us puzzlingly numb. But regardless of the reason, the outcome is usually the same. Emotional pain, like a persistent storm cloud, will just keep gathering above our heads, gaining more power until it finally unleashes its fury. There are a few reasons we avoid our pain.

Disconnection Driver 1: The Pain Paradox

As we've already seen, one factor that causes us to push away our pain is self-imposed pressure to tough things out. This response is often shaped by our belief that successful coping means unwavering composure—so anytime we buckle under the weight of our emotions, we feel like we're failing. While this view has recently been reinforced by the resilience movement, its origins date back millennia. Consider the ancient philosophy of Stoicism, which advocates self-control in uncontrollable situations.

Step 1: Probe Your Pain

The stoic concept of apatheia, or emotional tranquility, implies that overcoming our pain is simply a matter of will. As Marcus Aurelius once wrote, "If you are distressed by anything external, the pain is not due to the thing itself, but to your estimate of it; and this you have the power to revoke at any moment."

Modern psychologists call this strategy **emotion suppression.** And while it may make us seem strong on the outside, it carries big internal costs, like pulling our focus away from what really matters, draining our mental energy, messing with our mood, and increasing depression. In fact, research suggests that people who routinely suppress their feelings actually report *more* negative emotions. That's because, like missed interest payments on a loan, the more we suppress our feelings, the more they compound—and these "**negativity rebounds**" make us feel even worse than we did in the first place.

Despite its drawbacks, emotion suppression is usually the socially acceptable choice. In one survey I conducted with roughly 250 client employees at a major bank, two-thirds admitted that they were more likely to adopt a "grit and bear it" strategy than to face their emotions amid ongoing stress and crises. Another example comes from my friend who oversees leadership development at a multinational corporation in the Middle East. When he asked his executives to rate their own mental toughness—a trait centered on pushing through pain—most proudly labeled themselves a perfect ten out of ten. They were genuinely confused when my friend explained that this shouldn't be their standard![1]

Even if we suspect that powering through our pain isn't doing us any favors, we still might persist out of fear of the alternative. Yet while denial—like Jan announcing through clenched teeth, "Don't worry, I'm *just fine*"—may offer temporary relief, it only prolongs and intensifies our suffering.[2] I call this curious phenomenon, where we feel more pain in

---

[1]. If you're wondering what the "right" answer is, I'd offer that a score of about six or seven is likely much more beneficial because it allows us to know, and respond, when we are actually *not* okay.

[2]. Carl Jung would likely also agree. In his later years, he was a proponent of what he called "creative depression," where emotion was examined rather than stifled or survived (Deirdre Bair, *Jung: A Biography* [New York: Back Bay Books, 2004]).

the long term by avoiding it in the short term, **the pain paradox.** Take Jan, who had been enduring Andrew's meltdowns for months, pressing on, stoic and frozen. All the while, one thought loomed heavily: *I'm never going to crawl out of this.*

Disconnection Driver 2: Toxic Positivity

Have you ever taken the risk of sharing your pain with someone, only to immediately hear, "Stay positive!" "Don't worry, you'll be fine!" or "Everything happens for a reason!" I don't know about you, but these responses usually push me deeper into silent suffering. Generally, whenever we deviate from society's "I'm okay, you're okay" script, it can make others uncomfortable, as if we've committed a social violation. In this awkward space, well-meaning people may pressure us to reframe our situation more positively. This, as therapist Whitney Goodman writes in her book of the same name, is called **toxic positivity.** A sibling of grit gaslighting, toxic positivity means pressuring someone into silence or compliance when they share negative experiences or emotions, usually by insisting that everything is, or will be, just fine. As Goodman explains, toxic positivity not only makes us feel like our emotions are "wrong," but pressures us to convince ourselves (and others) that everything is okay. Yet simply saying something is okay doesn't *make* it okay—which is why Goodman compares this strategy to putting a Band-Aid on a bullet wound.

It's important to note that toxic positivity usually doesn't come from a bad place; most people are just echoing what they've been taught to say and do. These social scripts can be so powerful that they overshadow even the best of intentions and relationships. Moreover, when virtually everyone is fighting so hard to protect their time, energy, and well-being, it's even more tempting to gloss over the hard stuff so people can move on with their day. As Goodman writes, "Happiness and positivity have become both a goal and an obligation," and somewhere, "being a 'positive person' [started meaning] you're a robot who has to see the good in everything."

Yet some might argue that by pointing out reasons to be positive, they're not denying others' emotions but rather helping them reframe their experiences. However, research suggests that reframing our perspective to make something less emotionally charged[3] might not be much better than suppressing our feelings. In one study, reappraising negative events didn't reduce negative emotions for most participants, and for those aged seventeen to nineteen, it actually *increased* them.[4] This suggests that simply changing our perspective isn't a foolproof strategy.

Toxic positivity can be especially harmful for people with a tendency to prioritize others' needs over their own. Yes, my fellow people pleasers, we are even *more* likely to instinctively respond with, "You're right. It *will* be okay," to make the others feel comfortable and then silently vow never to speak up again. Take Jan, for example. The few times she tried to share that she was struggling, she was met with empty reassurances from her doctor to her friends and family. Eventually, too exhausted to argue, Jan stopped voicing her pain altogether.

I see at least two takeaways from the lesson that positivity isn't always positive. First, we don't need permission from other people to choose to pay attention to our pain. What's more, we must surround ourselves with those who have the courage to meet us where we are instead of where they want us to be (more in Chapter 11) and create space for *them* to safely share their emotional truths. After all, knowing what we now know, we should avoid being an agent of toxic positivity at all costs!

Disconnection Driver 3: Freeze-or-Faint

For decades, scientists believed that the human body's sole response to threat was the flight-or-fight reaction you read about in Chapter 1—where our autonomic nervous system kicks into gear and prepares us to either confront danger or swiftly exit the premises. However, psychiatrist

---

3. Careful readers of my book *Insight* might remember a discussion of the benefits of reframing in the context of improving our self-awareness. Rather than implying that reframing is never helpful, my point here is that when we're exhausted or under stress—that is, in "nonnormal" circumstances—reframing might be less effective than we think.

4. If you're scratching your head about this finding, this study's researchers were equally perplexed, suggesting the need for future research to untangle these effects.

Stephen Porges's groundbreaking work has challenged this narrative. Porges began questioning this conventional wisdom when he noticed that instead of reporting fight-or-flight sensations in response to stress and trauma, many of his patients described feeling physically immobilized, frozen, or completely detached from themselves.

Current theories didn't explain these responses, so Porges delved deeper. In 1994, he proposed the existence of a second neural circuit—even older than fight-or-flight—a **freeze-or-faint system** that takes over when we perceive extreme danger but can't escape or fight back. This system aims to conserve energy, minimize pain, and make us less appealing to predators by releasing pain-numbing endorphins and lowering our heart rate, blood pressure, and body temperature. The result is total physical and emotional shutdown: disassociation,[5] freezing, or fainting like a mouse feigning death in the jaws of a cat.

These findings became the basis of Porges's influential polyvagal theory, which goes on to explain that we don't get to choose between a mobilizing (fight-or-flight) or immobilizing (freeze-or-faint) response. Instead, our bodies decide based on their instinctive interpretation of the threat. Triggers can vary from person to person, but larger threats—especially those that we feel helpless against—are more likely to set off this system (more on triggers coming in Chapter 6). And just like fight-or-flight, our freeze-or-faint response can kick in when we least expect it. Recall from Chapter 1 that our bodies don't distinguish social or emotional threats from deadly physical ones. So when our brain registers extreme psychological stress, it decides we're in mortal danger—even if we're not aware of that stress. Porges offers the example of once fainting in an MRI machine despite feeling no conscious fear. Indeed, to involuntarily become completely emotionally numb is a confusing experience.

---

5. When we disassociate, we lose continuity between our thoughts, surroundings, and actions—it might look or feel like daydreaming, or it might be so disruptive that it results in a loss of identity or challenges with relationships and daily functioning (Stephen W. Porges, *The Pocket Guide to the Polyvagal Theory: The Transformative Power of Feeling Safe*, Interpersonal Neurobiology [New York: W. W Norton, 2017], 11).

In total, these disconnection drivers—internal pressure to suppress our emotions, toxic positivity from others, and involuntary freeze-or-faint responses—reveal the intricate ways our bodies and world steer us away from pain. (If you're curious how much you're avoiding your emotions, check out page 88 at the end of this chapter.)

Avoidance can often feel like the lesser of two evils: at least when we're not consciously feeling pain, we won't have to confront it and potentially fall apart. **What often eludes us is the option of taking a genuine interest in what our pain is trying to tell us.** While this might sound scary, it can actually reduce negative emotions and foster more positive ones in the mid- to long term. In one study, participants who spent just eight minutes reflecting on their reactions after a distressing event felt slightly worse than a control group immediately afterward but fared far better the next day; two days later, they were less affected by the event. Other research has revealed that these benefits last several months. So what if, instead of pushing through or pushing down our pain, we learned to harness it as valuable data to design a better life?

PAIN AS A SOURCE OF TRUTH

Lord Byron once noted that "adversity is the first path to truth." We've already seen that tough times offer unique opportunities for self-discovery, but where exactly in adversity do these truths lie?

In analyzing our shatterproof interviews, our team noticed that "the truth" emerged only when participants chose to pay attention to their pain. When we finally recognized this pattern, I kicked myself for not connecting the dots sooner. Over two decades of coaching executives, I've learned that transformations occur only when my clients confront and actively reject the status quo. Whether you're an ambitious professional stuck in an unfulfilling job, a new parent navigating marital problems, an entrepreneur struggling to revive a stagnant business, or anything else, one thing is certain: to improve your situation, you first have to take a long, hard look at what exactly is going wrong. (I often describe this process to my clients as a short fire walk, because, well, it's gonna hurt for a minute.)

More broadly, while both physical and emotional pain are unpleasant, they exist for a reason: pain is crucial for our survival. Indeed, most modern anguish is thought to exist because it once helped to ensure our ancestors' survival or reproduction. While physical pain signals bodily injury or harm, emotional pain indicates an unmet psychological need.

Imagine for a moment that you were born without the ability to feel any pain. Sixty-five-year-old Jo Cameron is one of only two people in the world known to carry a rare genetic mutation that renders her impervious to all its forms. Everyday annoyances don't bother her, and she's never needed pain medication after major surgeries—or even during childbirth, an experience she calls "quite enjoyable." But Cameron's life isn't the utopia we might imagine.[6] She frequently burns her arms in the oven, alerted only by the smell of singed flesh. And it wasn't until she found herself physically unable to walk that she discovered her hip had been destroyed by arthritis. When bad things happen, Cameron doesn't feel anxiety, anger, or sadness; but she also doesn't feel the drive to pursue a true sense of joy or happiness.

Pain, as researchers Randolph Nesse and Jay Schulkin have written, "always seems like a problem, but usually . . . is part of the solution." Fundamentally, **pain is a signal to pay attention**. A potent warning system, it disrupts our routines, impairs our performance, and demands our focus—effectively proclaiming, "Hey, if you don't take notice, you're going to be in a world of trouble!" It isn't just a mistake to think we can ignore pain without consequences; it's self-deception. Because, as the saying goes, the body keeps score.

Pain also forces us to challenge our preconceptions. Neuroscientists have identified the brain's insular cortex as the epicenter for processing painful experiences *and* learning from them. In the absence of pain, our behavior patterns calcify, but in its presence, it rouses us from complacency, forcing a reassessment of the status quo. As one of our shatterproof interviewees described, "I decided that I was never going to let something

---

6. People like Cameron experience worsening damage to their skin and joints, deformities, mobility problems, and early death (Marwan N. Baliki and A. Vania Apkarian, "Nociception, Pain, Negative Moods, and Behavior Selection," *Neuron* 87, no. 3 (2015): 474–91).

like that take me down again." With the right approach, embracing our discomfort unearths game-changing insight that boosts our well-being.

Finally, pain provides a path toward change. As we'll learn in Chapter 6, psychological pain is almost always the result of an unmet psychological need. If we don't trace our pain to its external triggers and the frustrated needs behind them, we'll stay trapped in a cycle of suboptimal choices. But by bringing awareness to our pain, we can discover how we might be getting in our own way—in this sense, pain powers personal reinvention.

Paying attention to our suffering doesn't mean inviting or relishing it. Instead, it means compassionately accepting our emotions with the goal of finding new ways of fulfilling our needs.[7] We don't have to enjoy these feelings, but we have the power to recognize them for what they are and where they can take us. As my colleague Brené Brown notes in her powerful book *Rising Strong*, admitting that we are cracking or even breaking is what clears our path forward. People who "rise" from challenges are willing to "walk into [their] stories of hurt" by acknowledging the power of their emotions and confronting their own discomfort. **This is why shatterproof people don't see the experience of pain as a personal failure—they see it as a power source.**

Yet how can we consciously confront our pain when it feels like we're tumbling off a psychological cliff? And when our freeze-or-faint response system has us in its grip, *can* we truly hear what our pain is trying to tell us?

PAIN AS A SOURCE OF PURPOSE

A few nights after Andrew's epic outburst, Jan found herself perched on the stairs aimlessly devouring a pint of Häagen-Dazs. As she waited for

---

7. A process within this theory called integrative emotion regulation is a relevant concept here. It's defined as a person's ability to experience negative emotions, explore their sources, and use this exploration to better understand themselves—which is why it's associated with greater well-being and self-determined behavior. (See Richard M. Ryan et al., "Building a Science of Motivated Persons: Self-Determination Theory's Empirical Approach to Human Experience and the Regulation of Behavior," *Motivation Science* 7, no. 2 [2021]: 97.)

that night's meltdown to subside, the events of the past week tumbled through her mind—her son's crisis, his admission of wanting to die, the haunting image of a trembling Ainsley in their freezing front yard.

Agony coursed through her like a long-awaited monsoon. The resilience that had gotten her to this point now fully depleted, Jan could no longer summon the energy to pretend. Yet in finally confronting the depth of her despair, a surge of defiance washed over her. With blinding clarity, something became obvious: her family could not go on like this. Not for one more day. Suppressing her feelings, and letting others dictate how she was supposed to feel, had prolonged and intensified her suffering for far too long. There *had* to be a better way, and she wouldn't rest until she discovered it. It didn't matter how many doctors, family members, or friends waved away her concerns. Her children would get the help they needed.

The next day, they turned the page. First, Jan and David gathered their kids for a heart-to-heart, reassuring them that their emotional chaos was real and wasn't their fault. Next, she rang their grit-gaslighting physician, demanding a referral to a general psychiatrist who could refer them to the specialists they needed. When the doctor attempted to downplay her concerns, she interrupted, "Let me be clear. I won't stop calling you until I get a referral." Sensing that Jan's new resolve would become an even bigger hassle for him, he relented.

This marked the beginning of Jan's "warrior mom" journey. A month later, the family had connected with a new doctor who understood the intricacies of Andrew's situation. He'd prescribed an antipsychotic medication, and to Jan's relief, the meltdowns miraculously subsided within days. But Jan didn't stop there. She delved into the research, collaborated with experts, and connected with other parents facing similar challenges. Her advocacy was relentless; she was driven by determination to identify the right providers and obtain accurate diagnoses for Andrew. No longer would she let others steer this ship—she was finally at the wheel.

Of course, none of this was easy. Confronting the truth about Andrew's condition was a painful process, full of grief and mourning for the life she'd envisioned for him. But facing this reality allowed Jan to accept that his path would be a different one. Despite moments of exhaustion and the occasional urge to disappear, she learned to recalibrate her expectations and made kindness to herself a mantra. As a couple, she and David learned to extend the benefit of the doubt to each other as they navigated their new reality.

Once the acute phase of the crisis subsided, Jan felt a compelling need to pay it forward. Helping others became her prescription for purpose. Decades later, she still serves on neurodiversity and mental health boards and advisory councils across North America, and her book, *Hold on Tight: A Parent's Journey Raising Children with Mental Illness*, has guided many others in navigating these experiences. Today, thirty-seven-year-old Andrew is a technology service technician, and thirty-five-year-old Ainsley is a child and youth counselor. Touching a necklace with the block letters STRENGTH, given to her by Ainsley, proud mom Jan gushes, "They are my heroes."

Jan's journey from fear and numbness to power and resolve is living proof of what can happen when we choose to trust what our pain is telling us—even when others make us question it. As Jan puts it, "You can't gain control of the situation and find solutions until you face your pain head-on. You have to accept the reality of your life and understand that it's okay, and that it's not going to be what you planned. On the other hand, you have to say, 'I'm going to kick the situation in its you-know-what.'"

PAIN AS A SOURCE OF DATA

In so many ways, Jan's emotional arc mirrored mine. Following the wedding weekend and my colleague's hurtful remark, I had hit my resilience ceiling. This triggered several months in freeze-or-faint mode while my life as I knew it collapsed around me. Persistent attempts to ignore my mounting anxiety and soldier on as if things were normal were undeniably failing. If anything, my refusal to acknowledge the emotional toll of my physical illness was just making things worse.

On one rather unremarkable day, I wondered, *If I continue this way, what will be the best- and worst-case outcomes?* The best case involved my symptoms magically disappearing, a possibility that felt laughable. The seemingly inevitable worst case was that I'd keep getting worse until something truly terrible happened. Clearly, "keeping calm and carrying on" was no longer an option.

It was time to shift my approach. I had to stop treating my feelings of frustration and despair as an enemy to defend myself against and start treating them as valuable signals. But I didn't just have to face my emotional distress; I had to take a closer look at my physical pain too. I opened a Word document and began brain dumping all of my symptoms. For each, I provided a brief history: When did I first notice it, and how had it progressed? Then I listed all the specialists I'd seen and their conclusions. One week and twenty-five single-spaced pages later, I had my first comprehensive medical history—and it hadn't required a single trip to a doctor's office. At this point, I noticed two things: first, a surge of motivation to take a more active role in my medical journey and then, a feeling of power.

Next came the reckoning with my emotional pain, which I knew would be much harder. One rainy afternoon, panic was setting in as I raced to clear out my inbox before my afternoon brain fog descended. Noticing my mounting anxiety, I closed my laptop, leaned back on my pillow, and took a deep breath. *Maybe,* I thought, *this is a good time to check in with my emotions.* Perhaps I could chat them up in a detached yet curious way, like a dinner party conversation with an out-of-town friend of a friend—nothing *too* deep, but enough to get to know them a little. "Is this your first time in the area?" "How long are you visiting?" "What are you doing during your visit?"

First things first: **How long had my painful emotions been visiting?** That was easy: since January 2021. The second question—**What were they doing (to me) during their visit?**—proved less straightforward. Grabbing my trusty Post-its, I did my best "feeling words" brainstorm (thank you, therapy). "Confused. Afraid. Scared. Unheard. Unseen. Victimized. Helpless. CRAZY." After reading the words back, I scrawled an angry afterthought: "THIS IS NOT ME."

The words described what I'd been feeling quite well. But they were completely at odds with how I saw myself and how I knew most people saw me. I was *trying* to hear what my pain was telling me, but its message seemed off. Then I realized I'd skipped the first question: **Was this my pain's first visit?** A sinking suspicion suggested the answer was no. Overwhelmed, I decided to take a nap and let all these new data percolate. Eventually, it hit me. My pain was trying to tell me that I had lost control of my life—and it was high time I took some steps to start claiming it back.

TOOLS TO OVERCOME EMOTION AVOIDANCE

According to polyvagal theory, our environment can activate one of three nervous system circuits. We've already learned about the first two—our mobilization (fight-or-flight) and immobilization (freeze-or-faint) systems—which kick in when we perceive danger. The third, our safety system, engages in the presence of cues, like a positive interaction with someone we care about, that help us feel safe and connected. And beyond the safety system's role in functions like reproduction, nursing, sleep, and digestion, it is essential for creative and generative thinking. When it's engaged, we can access the resources needed for behavior change. But we can't feel safe until we can find a way to escape our numb, emotionally disconnected state.

Porges offers several practical suggestions to help unfreeze ourselves. First and foremost, we can **forgive our bodies for their automatic reactions to threats.** Even though these involuntary reactions can be limiting, it's useful to remember that they don't exist to harm us; on the contrary, they are our body's way of helping us survive. So next time you're fighting, fleeing, or freezing, try gently saying, "Thank you, body, for trying to keep me safe. This isn't helping me right now, but I appreciate you making sure I survive in the best way you know how."

When I sit my coaching clients down to review the results of dozens of exhaustive interviews on how they're showing up as a leader, I often witness their involuntary autonomic reactions. I've seen every fight, flight, and freeze response you can imagine—crying, yelling, accusations of falsifying results, hiding in the bathroom, even punching a wall. While unpleasant, these are perfectly understandable responses to a lack of safety. Notably, none of these initial reactions have ever prevented a successful transformation. So whatever our knee-jerk reactions to threats may be, we deserve to cut ourselves some slack.

When we're gripped by fight-or-flight or freeze-or-faint responses, Porges's second suggestion is to **take charge of our personal narrative.** Some questions to explore: How does my history explain my physical responses? How do my reactions impact how I see myself? Do I view myself as a victim or a victor? Because the words we use matter, and we have the power to turn the page. In one study, when participants who tended to avoid their emotions were shown empowering words like *choice* and *opportunity*, they reported greater energy and well-being than those who read disempowering words like *must* and *should*. Simply acknowledging our agency over our circumstances, or even writing down and reflecting on empowering words, can make us less defensive, more self-respecting, and more able to heal. Porges encouragingly notes, "At some point, [our response] becomes our choice."

Porges's other scientifically supported methods of reprograming our nervous system for safety include pursuing positive social interactions, spending time with our favorite people or pets, giving and getting warmth and understanding, practicing self-compassion,[8] singing, and listening to lullabies or folk music. For each of these activities, he explains, "feeling safe *is* the treatment."

Once we've activated our safety system, we then can face our emotions with the right tools. The first is the **mood release,** a technique for diffusing acute negative emotions and experiences. Whenever the urge to suppress or avoid emotions arises, take a few minutes to articulate your thoughts and feelings, even if it feels temporarily intense. Express what's on your mind by saying or writing, "Right now, I am thinking . . ." and "Right now, I am feeling . . ." Research has shown that the very act of

---

8. For more practical tips on practicing self-compassion, see my book *Insight* or anything written by Kristin Neff.

Step 1: Probe Your Pain

putting our experiences into words gives us greater awareness and control over them.

The second tool, the **three-minute mood map,** helps form an important habit: noticing patterns in your feelings over time—specifically, by rating how often you experienced various emotions over the past week (to see this tool in its entirety, go to page 87). The third tool comes from acceptance and commitment therapy (ACT), a form of cognitive behavioral therapy that views confronting psychological pain as necessary to healing. One relevant ACT technique is called **leaves on a stream.**[9] Simply find a quiet space, close your eyes, and picture yourself sitting in front of a gently flowing stream. As negative thoughts or emotions come into your mind, imagine placing them on a leaf and watching them float away from you.[10] I personally love this tool; it has helped me learn to accept my feelings while minimizing my attachment to them. Once painful emotions become entities that we can name without guilt or judgment, we start to take away their power.

By following the Shatterproof Road Map, you will soon discover something game-changing: that **pain is not a personal failing; rather, it's a signal that one or more of our key psychological needs aren't being met.** Read on to discover how to uncover those unmet needs and why they undermine our happiness and success.

---

9. This practice was developed by Stephen Hayes, the father of ACT.

10. For a guided meditation of this exercise, check out this video from Milk & Honey Mental Health, "Leaves on a Stream," posted on YouTube, January 26, 2021, https://www.youtube.com/watch?v=exLaebgFO_8.

CHAPTER 5: KEY TAKEAWAYS AND TOOLS

"Pain always seems like a problem, but usually . . . is part of the solution."

- **The drivers of disconnection** (see the self-assessment at the end of this section):
  - **The pain paradox**: Deciding to ignore or push through our pain, which may offer temporary relief but prolongs our suffering over time.
  - **Toxic positivity**: Pressure from others to reframe our pain in a positive light.
  - **Freeze-or-faint**: An involuntary physical response that leaves us immobilized, frozen, and detached.
- **Pain is a power source**: Shatterproof people don't see their pain as a personal failure; it's a signal to pay attention, a chance to challenge our preconceptions, and a guide for finding new ways to meet our needs.
- **Tools to probe your pain**:
  - **Engage your safety system**: (1) Forgive your body for protecting you the best way it knows how. (2) Examine the words you're using in your personal narrative (victim or victor?). (3) Practice positive social interactions, giving and getting support, self-compassion, singing, and listening to lullabies or folk music.
  - **Befriend your pain**: How long have your emotions been visiting? What are they doing during their visit? Is this their first visit?
  - **Mood release**: Complete the following sentences: "Right now I am thinking . . ." and "Right now I am feeling . . ."
  - **Leaves on a stream**: Picture yourself sitting beside a gently flowing stream. As negative thoughts/emotions arise, imagine placing them on a leaf and watching them float away.
  - **The three-minute mood map**: Complete the following exercise weekly.

Step 1: Probe Your Pain

1	2	3	4
Less	The same	More	Much more

Think about your life right now. Compared to your "typical" self, please rate how much more or less you experienced each of the following this past week:

1. Sad[11]
2. Fearful
3. Disgusted
4. Angry
5. Surprised
6. Upset
7. Distressed
8. Jittery
9. Nervous
10. Ashamed
11. Guilty
12. Irritable
13. Hostile
14. Helpless
15. Hopeless
16. Lonely
17. Anxious
18. Depressed

Average Score	What It Means
2.0 or less	**LOW RISK**: Your responses suggest you're experiencing fewer negative emotions than usual. Continue monitoring for any changes, and if desired, continue to chapter 6 to proactively learn what to do.
2.1–3.0	**MODERATE RISK**: Your responses suggest you're experiencing slightly more negative emotions than usual. While you might not be in significant distress, keep an eye on this. To see what to do next, continue to chapter 6.
3.1 or more	**HIGH RISK**: Your responses suggest a higher level of emotional distress than usual. Use this opportunity to approach your pain with purpose by continuing to chapter 6.

---

11. Sad, fearful, disgusted, angry, surprised, and happy are the six universal emotions identified by Paul Ekman in "Facial Expressions," *Handbook of Cognition and Emotion* 16, no. 301 (1999): e320.

Self-Assessment: How Much Are You Avoiding Your Emotions?

This exercise can be especially helpful when you're feeling particularly down.

Consider your emotions lately and put a check mark next to the items that describe how you're feeling.

1. I feel frozen.
2. I feel numb.
3. I feel like I'm not in my own body.
4. I feel like I'm barely keeping it together.
5. I feel exhausted, irritable, or isolated.
6. I am spending energy convincing myself and others that I'm fine.
7. Other people are frustrated that I can't cope as well as I should.
8. Other people don't know why I can't be more positive.
9. I feel like I'm annoying my friends and family with my problems.

Items	What It Means
1, 2, 3	The more of these items you checked, the more likely you are to be experiencing a **freeze-or-faint** response.
4, 5, 6	The more of these items you checked, the more likely you are to be experiencing **the pain paradox,** where you push through to avoid breaking, but experience more negative emotions in the mid- to long term.
7, 8, 9	The more of these items you checked, the more likely you are to be experiencing external pressure from others to push through, like **toxic positivity.**

Chapter 6

Step 2: Trace Your Triggers

> "Everything that . . . happens externally is occurring . . .
> to trigger something within us—to expand
> us and take us back to who we truly are."
> —Anita Moorjani

Emotional distress is almost always triggered by *something*, even when we don't recognize it at the time.

In Ross's case, that something was Carson, the new junior realtor at the real estate agency where he worked. As one of the most senior employees, Ross was tasked with teaching Carson, who also happened to be the boss's nephew, the ropes. Despite getting along with almost everyone, Ross immediately felt frustrated—offended, even—by his interactions with his newest colleague.

Carson had many opinions and few qualms about voicing them. On his second day, he questioned the team's structure, a stance Ross felt compelled to defend. He then went on to challenge *everything* Ross taught him, meeting each process with endless questions or outright criticism. And it didn't end there. An excessive socializer with an appetite for gossip, Carson's commentary on his colleagues' lives was more editorial than supportive—like the time he eagerly probed Ross's devout religious beliefs as if disputing them outright. The first time it happened, Ross was so enraged that he literally had to leave the building.

Yet for reasons Ross couldn't begin to fathom, everyone else seemed to *love* Carson.

As the months dragged on, Carson turned Ross's once-sacred workplace into a minefield. Frustration lurked around every corner, and each interaction churned the seasoned realtor's stomach. As Ross's simmering rage festered, for the first time ever, he noticed a nagging insecurity about his performance and standing in the company. To discharge the tension, he would vent to anyone who'd listen, but his coworkers mostly responded with blank stares. At home, refusing to utter Carson's name outside working hours, Ross nicknamed his young colleague "The Deskpest"—a moniker that Ross's wife of twenty years loyally adopted.

Meanwhile, Carson kept climbing the ranks, eventually landing the position of associate broker. *Associate broker?!* Ross seethed. *That child couldn't broker his way out of a paper bag!* Then one day, Ross hit his breaking point. During a team meeting, he'd just unveiled a meticulously crafted plan for his new high-profile listing. The room buzzed with excitement, and Ross knew he'd killed it. Then, his eyes landed on Carson, frowning with his arms crossed.

Ross resisted the urge to charge across the table.

"Well," Carson began, "it's a solid plan. An oldie but goodie, am I right?" Looking around, he asked, "Does anyone have any ideas for how we can leverage more social media marketing?"

The thinly veiled criticism hung in the air. Face flushing, Ross rose so abruptly that he accidentally knocked back his chair. His typically measured voice trembling with incandescent rage, he announced, "Enough... is... ENOUGH!" Turning on his heel, he stormed out of the conference room as his colleagues' mouths collectively fell to the floor.

When Ross arrived home, his wife knew immediately that something was wrong. "The Deskpest again?" she asked. Ross nodded, his face twisting into a scowl. "Gosh, honey, to think of the number of good days he's stolen from you. And for what?!" They went inside and started dinner, but her comment lingered. He couldn't remember a time he'd felt so angry. What was it about Carson that triggered such an intense emotional response?

In Ernest Hemingway's masterpiece *The Sun Also Rises*, war veteran Mike Campbell is asked how he went bankrupt. "'Two ways," he quips.

Step 2: Trace Your Triggers

"Gradually, and then suddenly.'" Likewise, Ross's breaking point wasn't as sudden as it felt. Carson had been grating on his nerves and amping up his anxiety since his arrival—the casually ageist comment at the team meeting was just the latest in a long string of triggering interactions, "suddenly" sending Ross over the edge he'd been inching toward for months.

THE MISSING LINK: THREE-TO-THRIVE NEEDS

The year 2005 marked Tom Cruise's infamous couch jump, the founding of YouTube, the disbanding of Destiny's Child, and a milestone in my academic journey. As a third-year PhD student in industrial/organizational psychology, I embarked on my "comps," an intimidating, multi-year endeavor involving deep dives into one's field, churning out original research for one's faculty to dissect, and (hopefully) passing a grueling two-day exam.

As stressful as that experience was, I found an unexpected silver lining: an excuse to read all the empirical journal articles I could get my hands on (nerd alert!). One in particular caught my interest. Its architects, social psychologist Edward Deci and clinical psychologist Richard Ryan, first met in 1977 at the University of Rochester where they forged a lifelong collaboration that produced one of psychology's most influential theories.[1]

I vividly remember reading about self-determination theory (SDT) for the first time. One hot summer day, I sat in my cramped student office surrounded by piles of academic papers. SDT, one article began, reveals the defining factors that foster human flourishing. When I read that word—*flourishing*—I felt a surge of excitement, as if a light had turned on. Deci and Ryan's work left an indelible mark on my twenty-four-year-old mind, and the SDT lens still informs my work to this day. While many researchers at the time were casting humans as innately

---

1. In fact, it's not just a theory—it's considered a *meta*-theory (an overarching theory with six individual theories under it).

reactive to our environment, Deci and Ryan saw a hardwired motivation to shape and optimize it. While we can become "apathetic, alienated, and irresponsible," they wrote, at our core, we are growth-oriented, "curious, vital, and self-motivated." The million-dollar question: **What brings out the "best" rather than the "beast" in us?**

The answer turned out to be the clearest road map to a better life I'd ever come across (decades later, that's still true). SDT identifies three universal needs that human beings are biologically programmed to seek, and which provide a direct path to fulfillment, motivation, growth, and self-actualization—I call them the **three-to-thrive needs**.[2] When any of these needs go unmet, however, we become susceptible to understandable but ultimately unhelpful behaviors like reactivity, defensiveness, and other patterns that make flourishing virtually impossible. In other words, when we experience three-to-thrive **need thwarting** (also known as need frustration), we are prone to behaviors like the ones Ross exhibited around Carson.[3]

The first three-to-thrive need is **confidence:**[4] the belief that we're effective in our actions, capable of achieving our goals, and able to grow and learn new things. Confidence—about our problem-solving at work, parenting at home, or mastery of a challenging hobby—brings feelings of worthiness, pride, learning, growth, and self-acceptance. This intrinsic hunger for confidence is rooted in our evolutionary history. The psychological rewards our ancestors got from exploring, learning, and performing at their best played a crucial role in escaping danger, exploring new places and food sources, and following tribal rituals.

---

2. Research has shown that three-to-thrive needs are important for well-being above and beyond the need for safety, and are just as important regardless of how fulfilled (or not) our need for safety is (Beiwen Chen et al., "Does Psychological Need Satisfaction Matter When Environmental or Financial Safety Are at Risk?," *Journal of Happiness Studies* 16, no. 3 [2015]: 745–66).

3. Notably, the concept of need satisfaction has yet to be discussed in resilience research. This is more evidence that becoming shatterproof is indeed a distinct approach from resilience. (See Daniel J. Brown, Mustafa Sarkar, and Karen Howells, "Growth, Resilience, and Thriving: A Jangle Fallacy?," in *Growth Following Adversity in Sport: A Mechanism to Positive Change*, ed. Ross Wadey, Melissa Day, and Karen Howells [New York: Routledge, 2020], 59–72).

4. While Deci and Ryan use slightly different terms for the needs—competence (confidence), autonomy (choice), and belonging (connection)—I prefer my own for alliteration and simplicity.

The second three-to-thrive need, **choice,** means feeling free to function without pressure or threat, acting with agency and integrity, and staying true to ourselves. Whether it's through selecting a fulfilling career, leaving a toxic relationship, or advocating for a meaningful cause, choice helps us feel authentic, aware, and aligned as well as purposeful, empowered, and engaged. Choice also has evolutionary roots. An innate desire to chart our own destiny helped early humans evade coercive forces, like an enemy tribe, that could jeopardize their well-being or lead them down perilous paths. Similarly, when our choices are driven by our desires and values (i.e., motivation) rather than external pressure (i.e., mustivation), we forge a life of purpose and meaning.

The final need is **connection,** the sense that we belong, get along with others, and experience mutual closeness and support. Whether found in the comradery of your pickup basketball team, the support of your best friend at work, or feeling "seen" by your spouse, connection helps us feel included, valued, cared for, appreciated, encouraged, comforted, and championed. It's an innate human desire because early humans' survival odds greatly improved by cooperating for hunting, gathering, and protection. Today, countless studies reveal that social connection is not optional for our mental and physical well-being; in fact, even people who report finding close relationships unimportant benefit from improved connection.

Fundamentally, **confidence keeps us growing, choice keeps us authentic, and connection keeps us together.** Research shows that need satisfaction boosts well-being, life satisfaction, and performance. It helps us stick to healthy habits, stay engaged, and grow through adversity. Three-to-thrive need satisfaction also makes us more self-aware, empathetic, and cool under pressure. It improves our relationships, both romantic and platonic. At school, need satisfaction enhances classroom experiences while reducing conflict and bullying. Fulfilled needs even change our brains, activating reward centers, supercharging motivation, and elevating decision-making.

On the flip side, when our environment frustrates our needs, the negative consequences are immediate. In Ross's case, Carson managed to effortlessly thwart all three: his constant criticism crippled Ross's confidence, being forced to train Carson undermined his choice, and his colleagues' indifference to these slights strained his connections. All of this (as research shows) left him emotionally and physically drained—burned out, even—and more vulnerable to judgmental and aggressive behavior. Unmet needs also heighten anxiety, depression,[5] cynicism, and existential loneliness, preventing us from fulfilling our potential. And when need frustration begins early in life, it can also predict antisocial behavior, borderline personality disorder, and, in rarer cases, violence and murder.

These findings underscore why **it's crucial to move beyond merely managing our emotional reactions and learn to identify the unmet needs beneath them.** Consider John, the CEO of a fast-growing analytics firm I coached, whose pattern of frustration and reactive behavior in meetings was sucking the psychological safety right out of the room. John wondered, "What's my problem? Why do I keep doing this?"

Here, a quick fix would have been merely teaching him to better control his reactions, but diving deeper into the issue causing them proved a far bigger opportunity. To borrow from my friend Dan Heath's excellent book *Upstream*, we had to identify the root problem rather than reactively treat the symptoms (which, in this case, happened to be John's reactivity). The root cause turned out to be a series of process issues creating missed milestones, which threatened John's confidence in his own effectiveness, and drove him to lash out at his team.

"John," I told him, "you asked me what your problem was. Your problem is that you're a human being whose needs aren't being met." Once we discovered the need-based patterns triggering his reactivity, he was able to tackle the team's process problems, leading to fewer missed milestones and a more trusting team dynamic that was no longer being stifled by the CEO. Meanwhile, John felt more confident and, as a result, fostered more safety in his team.

---

5. Stats nerds: One study found a stunning correlation between need thwarting and depression across cultures—a whopping .60! (Beiwen Chen et al., "Basic Psychological Need Satisfaction, Need Frustration, and Need Strength across Four Cultures," *Motivation and Emotion* 39 [2015]: 216–36).

Step 2: Trace Your Triggers

DEMYSTIFYING TRIGGERS

Triggers are the tipping points that instantly flip our switch from "okay" to "not okay." In an era of widespread "trigger warnings," the term is probably familiar to you. In such definitions, a trigger is seemingly anything that might leave us feeling upset, hurt, or offended. This popular understanding can be markedly different from the definition most therapists use, which is more closely tied to specific reminders of significant trauma, often for people with post-traumatic stress disorder (PTSD).[6]

Given these many interpretations, let me be clear: When I talk about triggers, I mean something very specific. In the context of becoming shatterproof, **triggers aren't simply things that upset us; they're signals or reminders of unmet needs.** And while three-to-thrive needs are universal, the triggers that upset those needs can be more personal. Shaped by our upbringing, personality, experiences, and expectations, they typically provoke involuntary, immediate, and often disproportionate emotional reactions in specific situations. In their presence, we become hypersensitive, have difficulty controlling our reactions, and act out of character.

Though triggers can cut deep, they're generally not inherently something to avoid.[7] In fact, avoiding triggers can rob us of important opportunities to explore our reactions and unmet needs. What's more, the effort it takes to avoid situations that *might* be triggering leaves us less able to process and learn from the emotions the trigger stirs up. Finally, avoidance strategies can be logistically problematic. As trauma therapist Carolyn Spring notes, "Triggers are like little psychic explosions that crash through avoidance," and because triggers are subjective, we often don't know what will set us off. Or, even if we're not actively avoiding

---

6. One of the most insightful things I've ever heard about trauma describes it as an injury to our nervous system that prevents us from feeling safe. The good news is that through various therapeutic techniques, trauma can be processed and managed.

7. If you have been diagnosed with PTSD or are exhibiting symptoms of trauma, make sure to run this advice by your therapist. Especially early in recovery, true trauma triggers can be a different beast and should be treated very carefully in general, including in the context of becoming shatterproof. (I say this from personal experience!) I recommend consulting with your mental health professional to determine the best strategy for you.

triggers, it can be easy to brush past them—as Ross did—simply believing they're crosses we're supposed to bear. But as we learned in Chapter 3, this mindset can eventually push our resilience to its limit, leaving us vulnerable to snapping at even the tiniest offense or stressor.

In some cases, triggers don't just signal *current* unmet needs but remind us of past ones as well. Many years ago, I started to notice my friends gradually showing up less and less to our cherished Thursday night happy hours, a tradition my husband at the time and several of our friends managed to maintain for fifteen years. Thursday night happy hour involves drinking discounted cocktails in interesting establishments around Denver while consuming all the appetizers we can stomach, alongside gossip and laughter and all the things that make life feel full and wonderful.

Historically, our friends have fiercely guarded this tradition, so when enthusiasm began to wane, let's just say that I wasn't thrilled. On weeks when happy hour was canceled, I'd feel a dull and desperate ache and then proceed to be grouchy and sullen all evening. I knew I was overreacting, but for some reason, I couldn't help myself.

One unusually tiring week, I was thrilled that Thursday night happy hour was *on*. Rob and Teresa were out of town, but all systems were go with Mike and Sue. Each day, my excitement built a little more. On Thursday morning, I bounded out of bed and started counting down the hours. Then, a few minutes before 5:00 p.m., we were waiting for the elevator in our building when a text message lit up my phone. It was from Sue.

"So sorry. We can't make it tonight. We have to go see a guy about a cat."

Feeling like I'd been punched in the stomach, I showed my phone to my husband. "Is she serious!? A cat? She *must* be joking." I texted Sue a long line of crying-laughing emojis, my heart thudding in my chest.

Several long seconds later, she replied: "We're adopting a cat! Sorry to bail! Love you."

"*Sorry to bail?*" I repeated to Dave, who rolled his eyes supportively. "Who is this busy cat that we all have to reorganize our lives around?" I seethed, "Does its schedule only permit a Thursday night appointment? Are its Fridays an absolute *nightmare*?!"

Step 2: Trace Your Triggers

Back at home, I proceeded to, well, completely lose my shit. I slammed my phone on the table, erupting in expletives as I stormed back to my closet and ripped off my going-out clothes—all the while shouting, "This friendship is OVER! O-V-E-R!"

You know, a totally proportionate reaction.

Convinced I was wisely protecting myself from future hurt, I didn't speak to Sue for months. (Ugh, I *know*!) Yet I couldn't help but notice that shunning such a close friend did little to improve my well-being. If anything, I felt worse. At work, I was less able to tolerate daily annoyances; at home, I was snapping at people more than I should have.

At one point, my friend Teresa offered to take me out for dinner, revealing her agenda only after our sky-high salads arrived at the table. "Soooooooo . . ." she said, "I was hoping we could talk about what's happening with you and Sue."

I sighed. "No need," I replied stubbornly. "I'm not forgiving her yet."

Teresa paused, taking a sip of her Malbec. Her kind eyes meeting mine, she ventured, "I'm not *exactly* sure what happened—all I know is that Sue canceled on you, and you haven't spoken to her since." I nodded righteously. "But do you think," she offered, "that maybe, just maybe, this is really about something else, and not about Sue at all?"

I'd never even considered this possibility (psychologist, heal thyself!). Wincing, I conceded, "Perhaps."

Happy with our progress, Teresa changed the subject and we finished our dinner.

As William Faulkner once observed, "The past is never dead. It's not even past." He and Teresa were right. Something about this trivial annoyance had deeply triggered something unresolved from long ago. As I walked home from dinner, I tried to remember how I felt when I read Sue's text message: unimportant, discarded, an afterthought (pain as data!). Then I asked myself, *When else have I felt like this?*[8] The answer came startlingly quickly.

When I was five, my dad left my mom and soon remarried. After a long, acrimonious divorce, my mom and stepdad won primary custody.

---

8. This is the "compare and contrast" tool from *Insight* in action.

And though my dad, stepmom, and two half-sisters did try to include me in family events, they often (I think) simply forgot. I'd frequently hear about their celebrations and vacations—a hike in the mountains, a big Father's Day dinner, an amazing trip to Mexico—after the fact. I'd see photos of smiling faces in front of beautiful landscapes under perfect blue skies, with all the love and happiness between them on full display, and I wasn't included in any of it. Worse, I had to listen to their stories and laugh along. My dad's family didn't realize it (and probably still doesn't), but they broke my heart over and over, for decades.

Recalling this experience brought a hard lump to my throat, which was all the evidence I needed. For the first time, I could clearly tie my outsized reaction to the disintegration of Thursday night happy hour to feeling excluded and unimportant throughout my childhood—a wound that cut so deep it was resurfacing decades later, in a seemingly unrelated situation.

It may be helpful to know that **when painful emotions from our past resurface, it's never just "in our head."** Reminders of past pain activate the neural circuits that store these memories—whether or not we are aware of it—creating a kind of conditioned threat response that can include intense anger, sadness, emotional outbursts, or physical symptoms like the lump in my throat I experienced. This is how a benign disappointment with a friend turned into a four-alarm abandonment fire. Thankfully, recognizing this pattern helped me put the situation in perspective. There was, of course, no logical connection between feeling brushed off as a child and Sue and Mike's last-minute cancelation. It was time to reconcile with Sue, which I did. I even got to meet her new cat, Ted, who kindly found time in his busy schedule to see me.

A few months later, the six of us (Sue, Mike, Teresa, Rob, Dave, and I) embarked on a trip of a lifetime to Italy—one that would never have materialized if I hadn't taken charge of my trigger.[9]

---

[9]. If you find yourself facing an old or extremely traumatizing trigger, and the methods we're about to cover don't help, I suggest looking into traumatic incident reduction (designed to help people process and reduce their reactions to past traumatic experiences), eye movement desensitization and reprocessing (or EMDR, designed to alleviate the distress associated with traumatic memories), or newer but promising approaches like brainspotting and emotional transformation therapy (ETT).

Step 2: Trace Your Triggers

TAKING CHARGE OF YOUR TRIGGERS

On February 9, 2018, inside the Gangneung Ice Arena on the outskirts of Pyeongchang, South Korea, American figure skater Nathan Chen is seconds away from his Olympic debut. Today, he'll tackle the team event, then singles a week later, where a short program and free skate will, according to commentators and fans, finally etch his name in the annals of Olympic history. The air in the arena is as tense as a bowstring about to release.

A natural-born talent, Nathan has been nimble on the ice since age three. His journey to the twenty-third Winter Olympic Games owes much to the tutelage of his tireless mother, Hetty Wang. An immigrant from Beijing, Hetty became a self-taught scholar in the art of training winners, meticulously studying sports textbooks and methods of the world's most successful athletes and coaches. In Nathan, she instilled the belief that "when you had work to do, nothing else mattered." She would regularly drive him from their home in Utah to train with a top coach in California; barely able to scrape together money to pay for his lessons, they'd sometimes sleep in their car.

And now, many years spent powering through have paid off. The two-time national champion arrives at the Pyeongchang Olympics as the only skater in active competition who can land five different quadruple jumps (the hardest move in skating). At this point in the season, Nathan has won every competition he's entered. Just last month, he placed first at the US national championships, landing *seven* quads and leaving his closest competitor an almost inconceivable fifty-five points in the dust. Today, billions are tuning in to see the eighteen-year-old—known by fans as the "Quad King"—earn his rightful spot on the Olympic podium.

Two minutes and forty seconds later, it's all over. The audience is stunned. And not in a good way.

"Honestly, it was bad," Nathan tells reporters. "I made as many mistakes as I possibly could have." The media agrees, calling it "perhaps the worst performance of his life."

But all is not lost. Next week, he'll have a chance to redeem himself in his singles short program. In interviews, he is determined: "Having the experience to do it once allows me to do it better the next time." And everyone loves a comeback.

The morning of February 16, Nathan skates onto the Olympic ice for the second time, to a chorus of cheers, applause, and hundreds of cameras clicking.

In his first jump, he falls. *Epically.*

Then he falls again.

By the time the music ends, the Quad King hasn't landed a single quad. Stepping off the ice, he can't even look at his coach. He finishes seventeenth out of twenty-four skaters. A *USA Today* reporter writes in disbelief, "Incredibly, Nathan Chen bombed again."

Nathan didn't merely crack at the 2018 Winter Olympic Games; he completely shattered. And not once, but twice. Of course, his uncharacteristically poor performance had almost nothing to do with the skater's ability. Something had obviously triggered Nathan when he stepped out onto the ice, causing him to lose focus and epically bungle his programs.[10] Clearly, Nathan noticed that something had thrown him off kilter. But the awareness that we're triggered isn't the same as knowing what exactly what has triggered us. We'll return to Nathan's story in Chapter 7, but for now, let's look at a two-step process for taking charge of our triggers before they take charge of us.

Step 1: Identify Patterns

The weekend after Ross's outburst, he rose before dawn and set out on a long, contemplative walk to figure out The Deskpest situation. Thinking back to times he'd felt triggered by Carson, he identified three consistent patterns. First, his inner monologue instantly went haywire. Normally affable and confident, his thoughts became paranoid and self-doubting in

---

[10]. If you're wondering how I know all this, it's because I read Nathan Chen's authentic autobiography, *One Jump at a Time*.

the presence of his young colleague. Second, his emotions intensified (he felt angry, offended, and aggrieved) alongside physical stress symptoms like his stomach dropping. Finally, his behavior changed on a dime, cringing as he recalled his Dr. Jekyll/Mr. Hyde conference room incident.

I can't promise that exploring your triggers will be particularly pleasant, but it *will* be valuable to reveal the situations that most derail you. Consider this: When you're stressed, overwhelmed, or exhausted, what sends *you* over the edge? Answers that spring to mind may include "not being taken seriously" (confidence), "being told what do to" (choice), or "being excluded" (connection). These are just a few of the most common triggers; you'll find a full list at the end of this chapter along with which need(s) they thwart. As you review the list, you'll probably notice that confidence triggers typically relate to our performance and contributions, choice triggers largely come from people, systems and random chance, and connection triggers usually stem from our interpersonal interactions. Generally, the objective facts of the situation matter less than our *subjective* interpretation of them. For example, an outside observer might see Ross and Carson as two conflicting personalities who both mean well, but to Ross, Carson was mistreating him—for all intents and purposes, that was true because it was how Ross felt.

Also, as a rule of thumb, situations often contain multiple triggers, and generally, more triggers mean more extreme reactions. Some triggers neatly align with one unmet need (e.g., rejection threatens connection), while others can encroach on two or more. Expectations, for example, threaten confidence and choice, and coercion threatens choice and connection. Indeed, once Ross recognized the many ways the Carson situation was triggering him (criticism, inferiority, threat, disregard, unfairness, conflict; see page 103)—it was no wonder the associate broker brought him to his breaking point!

Step 2: Trace Triggers to Unmet Needs

When we are open to learning from them, our triggers can be our teachers, revealing the unmet need(s) lurking beneath our outsized reactions.

But with so many factors at play, tracing triggers to our unmet needs can get complicated quickly. The good news is that we can keep things simple while still unearthing actionable insights with the **need audit,** a tool to nail down our *most* thwarted need (see Appendix C for the complete tool).

First, we can identify our fixations. Research suggests that when our needs aren't met, we become especially sensitive to anything that blocks or threatens them—and the more a need is frustrated, the more we fixate on it. In the absence of confidence, for example, we might zero in on our mistakes or ruminate about how others perceive our performance. Amid thwarted choice, we can latch onto signs that we're being manipulated or treated unfairly. When we're disconnected, we could become extra sensitive to signs that we're being excluded or are unloved. In general, **our greatest unmet need creates our most intense fixation.**

What's more, we can also explore our biggest fears. When we don't feel confident, our fears center around feeling unworthy, not good enough, inferior, or incapable. When our need for confidence is unmet, our fears center on unworthiness, inadequacy, or inferiority. Fears that we're actually powerless, a cog in the machine, or don't know who we are, are signs of thwarted choice. With disconnection, we fear that we are unlovable or unimportant. **Our greatest need usually fosters our most visceral fear.** For a shortcut, ask yourself: What thoughts are keeping me awake at night?

Through "a lot of reflection," Ross identified his biggest fixation (any evidence indicating he didn't know what he was doing) and fear (that he wasn't as valuable as he thought)—both of which suggest that confidence was his biggest unmet need. With that insight came the shocking likelihood that Carson wasn't actually trying to belittle him; he was just an imperfect person who wanted to do his job well too.

Ross returned home from his walk with a new perspective that would help him engage Carson in a healthier way. Fast-forward several months, and Ross now considers Carson a "work friend." They exchange fist-bump greetings, ask after each other's families, and have even colisted a few properties together—a collaboration that would have been inconceivable just months earlier. And all of it, without a single outburst.

Step 2: Trace Your Triggers

CHAPTER 6: KEY TAKEAWAYS AND TOOLS

*"If we don't take charge of triggers,
they'll take charge of us."*

- **Self-determination theory**: A theory outlining what brings out the "best" and the "beast" in us.
- **Three-to-thrive needs**: Needs that humans are biologically programmed to seek, and when met, help us flourish:
  - **Confidence**: A sense that we're doing well and getting better.
  - **Choice**: A sense of agency and authenticity.
  - **Connection**: A sense of belonging and mutual support.
- **Triggers**: Signals or reminders of unmet three-to-thrive needs that flip us from "okay" to "not okay" (see below for a list).
  - **How to know you're being triggered**: (1) Your thoughts and inner monologue turn negative. (2) Your emotions and physical symptoms worsen. (3) Your behavior becomes less well-controlled.
  - **Tracing current triggers to past ones**: Ask "When else have I felt like this?"
- **Need audit**: A tool to identify our *most* thwarted need by reflecting on our fears and fixations (see Appendix C).

Common Triggers

Primary Unmet Need	Trigger	Examples	Secondary Unmet Need(s)		
			CONF.	CHO.	CONN.
Confidence (Performance and contributions)	**Expectations**: Pressure to perform to a certain standard.	• Demands to achieve a particular result or outcome. • Overly demanding tasks or expectations (e.g., time pressure, job pressure, etc.). • Having others' approval or attention depending on our performance.		X	
	Monotony: Unchallenging or unrewarding tasks.	• Repetitive or boring tasks. • Not having enough to do. • Not making a meaningful contribution or doing what we're good at.		X	

Primary Unmet Need	Trigger	Examples	Secondary Unmet Need(s)		
			CONF.	CHO.	CONN.
	Chaos: Unclear or changing standards.	• Lack of clarity, certainty, or predictability. • Constant changes in roles, goals, or rules. • Unclear expectations or standards.	X		
	Setbacks: Falling short despite our best efforts.	• Falling short on important goals and outcomes. • Failing (especially when we tried hard and "should" have succeeded). • Not reaching the standards we've set for ourselves.	X		
	Criticism: Having our faults highlighted.	• Being judged, doubted, or criticized. • Having our shortcomings or weaknesses pointed out. • Others seeing us as ineffective or incompetent.			X
	Inferiority: Being treated as "less than."	• Not being as capable or qualified as those around us. • Not having our contributions acknowledged or appreciated. • Others believing they're superior to us (e.g., condescension, mansplaining, etc.).			X
Choice (People, systems, random chance)	**Suppression**: Internal or external pressure to act against our true selves.	• Pressure to go against our values, interests, or goals. • Pressure to do what we "should" instead of what we want. • Pressure to wear a social mask or hide our true feelings.			X
	Coercion: External factors forcing us to obey.	• Being forced to act against our will. • Being prevented from making choices or having our choices limited. • Being guilted into acting a certain way.			X
	Loss: Losing something that we value.	• Something important being taken away from us. • Losing control over our health, livelihood, or reputation. • Having to abandon a valued future.	X		X
	Disregard: Being minimized or invalidated.	• Having our opinions or experiences denied or devalued (i.e., gaslighting). • Not being listened to or heard or being interrupted. • Having our time wasted or not respected.	X		X

Step 2: Trace Your Triggers

Primary Unmet Need	Trigger	Examples	Secondary Unmet Need(s)		
			CONF.	CHO.	CONN.
Connection (Interpersonal interactions)	**Unfairness**: Bias, inequity, or discrimination.	• Unfair outcomes or processes (like stealing credit). • Being treated differently due to personal characteristics (race, age, gender, sexuality, disability, etc.). • Favoritism, inconsistent standards, lack of accountability for others.	X		X
	Voicelessness: Lack of transparency or input.	• Being excluded from decisions that affect us. • Having our input or preferences discounted. • Being intentionally misled.			X
	Rejection: Being dismissed or abandoned.	• Being left out or ignored. • Being actively avoided or disliked. • Being cut off from our connections.		X	
	Neglect: Lack of care or concern.	• Being treated coldly, inattentively, or indifferently. • Lack of support or concern for others. • Others being favored over us.		X	
	Conflict: Interpersonal struggles or misunderstandings.	• Tension with important others. • Disagreements, arguments, or fights. • Having our intentions misunderstood or judged.	X	X	
	Cruelty: Dehumanizing or hurtful treatment.	• Being punished, blamed, humiliated, or sabotaged. • Abuse, microaggressions, or passive-aggressiveness. • Feeling objectified or used for someone else's gain.		X	
	Betrayal: Someone breaking our trust or loyalty.	• Someone we trust lying, being unfaithful, or duplicitous toward us. • Having our boundaries, wishes, or confidentiality breached. • Someone breaking a key commitment or putting their interests over ours.		X	

Chapter 7

Step 3: Spot Your Shadows

> "The brightest flame casts the darkest shadow."
> —George R. R. Martin

In the serene Swiss town of Kesswil, where the gentle breezes of Lake Constance caressed the foothills of the majestic Alps, a young boy began a solitary journey into his own mind. Born in 1875, the boy had a family lineage boasting philosophers, ministers, and doctors—including whispered ties to the illustrious Johann Wolfgang von Goethe.

An only child until the age of nine, his active mind, vibrant dreams, and precocious analysis of those around him revealed an intricate inner world that set him apart from his peers. Puzzled by their peculiar boy, his parents would leave him alone to play, giving life to his vivid imagination.

At the tender age of twelve, a fateful encounter triggered a retreat deeper into himself. Pushed onto a curbstone by another boy, he momentarily lost consciousness. Back at school, his mysterious fainting episodes became so distracting to fellow students that he was sent home for six months of solitary study. Thrilled with the chance to explore the world on his terms, he daydreamed, invented secret games, and escaped into the books in his father's library.

Upon his recovery, the boy embraced his studies with newfound seriousness, sometimes rising at 3:00 a.m. to pore over his lessons before setting out on a one-hour trek to school. By age thirteen, he was immersing himself in philosophy—Schopenhauer, St. Thomas Aquinas, Kant, Nietzsche. Ultimately, he chose to pursue the profession of psychiatry.

Step 3: Spot Your Shadows

In the early years of his career, he secured a modest position at a mental hospital, living a life dominated by work. That changed in 1901 with his betrothal to Emma Rauschenbach—a fellow scholar and Switzerland's second wealthiest heiress. Their union offered new financial security, emotional support, and energy to make a bigger impact.

Soon, the young psychiatrist was gaining an international reputation for his experiments on what our words reveal about our subconscious mind, eventually publishing a groundbreaking book on his findings. In the spring of 1906, he boldly sent his tome to Sigmund Freud, the (by then) famous founder of psychotherapy. Their initial encounter, a legendary thirteen-hour conversation at Freud's apartment, ignited a collaboration bound for combustion.

Despite Freud's desire for an intellectual heir, he found the young psychiatrist to be ill-suited for discipleship. Tension escalated when his mentee challenged the idea that sexual energy underpinned all psychological dysfunction—the central tenet underlying Freud's entire theory of psychoanalysis—and outwardly questioned why the theory's revered father couldn't seem to tame his own neuroses.

By late 1912, Freud announced an unconditional split, proclaiming he'd "lose nothing" from the termination of a collaboration that had "long been a thin thread; the lingering effect of past disappointments." In early 1913, the two men parted ways. This "failed bromance" plunged the young scholar into a profound crisis. Reeling from rejection, angst over his lost mentor, and uncertainty for his future, his emotions were so intense he wondered if he'd developed a psychological disorder.

It was at this point that the rudderless psychiatrist embarked on a self-imposed four-year exile. Refusing to succumb to his inner torment, he vowed instead to put it to use. He began intensely examining his dark fantasies, dreams, and beliefs, meticulously documenting the process in a series of aptly named "black books." At first, plumbing the depths of his subconscious frightened him—after all, he *was* a respected empirical scientist—yet he began to find the process "self-healing." Soon, he was testing his emerging methods on patients, many of whom had been treated unsuccessfully with Freud's psychoanalysis. He was further encouraged when they, too, noticed significant improvements.

Out of one man's crisis emerged a theory that remade the world's understanding of the human mind. You see, the once-solitary boy from Kesswil was none other than Carl Jung—one of history's most renowned and prolific psychologists.[1] And this dark period birthed one of Jung's most important contributions to his field.

As we navigate life, Jung's analytical theory explains, internalized social norms and values powerfully shape our ideas, emotions, and actions. For instance, we're taught in school to value academic excellence and behavioral discipline, building the belief that "successful" students get good grades and mind their teachers. But more often than we'd like, we violate these norms, giving in to our "shadows"—reservoirs of instinctive, norm-violating reactions we vehemently wish to avoid, like dark thoughts, self-destructive desires, and unpleasant qualities like rage, insecurity, arrogance, greed, and perfectionism.[2]

Most of the time, people manage to keep their shadows buried. It's only when we're triggered that, in defiance of our conscious mind, they rise to the surface, often instantly and involuntarily. In this chapter, you'll learn how to spot your **shadows**—those automatic responses to persistently thwarted needs that drive us to behave, as Jung put it, like "the person no one wants to be." The goal will be to set you up to replace your biggest self-limiting responses with shatterproof ones.

---

1. Jung, like many historical figures, was complicated. Many elements of his work, like his identification of psychological types (e.g., introversion and extraversion) and concepts like individuation, the collective unconscious, complexes, and synchronicity, are enduring fixtures in contemporary psychological thought. What's more, scientists are discovering that many of his observations about how consciousness operates mirror the actual physiology of the brain. At the same time, he suffered controversies ranging from his personal relationships to his treatment of women to allegations of anti-Semitism.

2. It's important to note that there can be genuinely evil parts to the shadow. But my guess is that you, dear reader, aren't dealing with uncontrollable urges for abuse, violence, or murder as much as the more "normal" day-to-day manifestations of the shadow, which is what we'll focus on in this chapter.

Step 3: Spot Your Shadows

INTO THE SHADOWS

When our confidence, choice, or connection needs are thwarted, we tend to search for "quick fixes" to restore a sense of self-worth, power, or approval. Rooted in our primal, instinctive drive for psychological survival, these **shadow goals** allow us to feel—at least temporarily—as though our needs are being satisfied. Yet they are, in fact, poor replacements for the authentic experience of confidence, choice, and connection. As we're about to see, **not all goals are created equal: some help us, some hurt us, and not all are consciously chosen.**

To better understand the concept of shadow goals, imagine coming home after an exhausting day. You're craving a healthy, delicious salad, but upon opening the fridge, you find your lettuce has spoiled. Frustrated, you seek a second choice; after all, you need to eat *something*. Scouring the freezer, you dig out a pizza. Tempted by its cheesy goodness, your stomach growls. *Forget being healthy. I WANT CARBS!*

In this analogy, the pizza represents your shadow goal—an adjacent alternative that's immediately satisfying, but unlike the salad, won't meet your body's need for a nutritious meal. (You can find a list of common shadow goals at the end of this chapter.)

Researchers believe that we subconsciously choose shadow goals to cope with the frustration, anxiety, and insecurity that unmet needs create. Shadow goals are shallow shortcuts,[3] as tempting as carbs smothered in cheese, to which we're most vulnerable when exhaustion is high and willpower is low. But no matter how much shadow goals seem to boost our mood or soothe our in-the-moment anxiety, they ultimately drain our energy, causing us to spin our wheels and lose sight of what we really need to thrive.

Shadow goals typically fulfill one of three **compensatory motives** (meaning, "If I can't have confidence/choice/connection, at least I can have *this*"). First, we may focus on **protecting** ourselves as an antidote from the guilt, shame, and bruised ego that frequently frustrated needs foster.

---

3. I would like to thank the astute beta reader for this gem!

For example, when we don't feel confident in our performance, we might become defensive in an attempt to fend off feelings of worthlessness or avoid judgment from others. When we feel helpless, we might try to reestablish control by rebelling or openly flouting expectations. When we feel unloved, we might be tempted to punish those who've hurt us. What protection motives have in common is that they externalize our problems, making them everyone's fault but our own.

Second, the motive of **proving** involves searching for any external evidence that—despite our fears to the contrary—we actually *are* worthy, powerful, and loved. For instance, if Tim faces constant criticism from his boss, he might work hard to achieve more than his colleagues or seek external rewards (like a raise, industry recognition, or promotion) to validate his competence to himself and others. To prove we're powerful, we might assert whatever authority we have by becoming bossy or controlling, often to people with less power, or by being overly controlling of ourselves. And to prove that we are liked and respected, we might go to great lengths to gain popularity or status.

The final compensatory motive, **preventing,** attempts to stop our momentary mood from worsening by escaping, ignoring, or downplaying issues. With frustrated confidence, for instance, we might opt out of situations where we feel like an impostor or avoid tasks we can't immediately master. When we lack choice, we can surrender to our constraints by giving up or giving in. And when we feel rejected, we may distance ourselves from others to avoid future hurt (as I did with my friend Sue in Chapter 6). Prevention can offer a short-term escape, but as we've seen throughout this book, the pain will eventually come back to bite us.

By paying attention to these three compensatory motives, we can better recognize the self-limiting patterns we're most prone to: Do you tend to be a protector, a prover, a preventer, or a little bit of each?

THE SCOURGE OF SHADOW GOALS

While shadow goals might seem (at least superficially) similar to three-to-thrive needs, our strategies for chasing them couldn't be more different.

Step 3: Spot Your Shadows

Psychologists distinguish between two basic human motivations: we can either do things out of a sense of authentic choice (**intrinsic motivation**) or out of pressure or expectation for rewards (**extrinsic motivation**). The former satisfies our three-to-thrive needs over time, while the latter actively distances us from them. Shadow goals are overwhelmingly extrinsic in nature, and their negative impact is profound. By causing us to unconsciously downplay, compromise, or replace our three-to-thrive needs with lesser alternatives, they trap us in a cycle of pressure, negativity, and ill-being. They even rewire our brains for less intrinsic motivation in the future. **Shadow goals, in other words, are paper tigers: poor replacements for the intrinsic experience of confidence, choice, and connection.**

Nathan Chen's Olympic journey, which you read about in the last chapter, is a vivid example. While he was no stranger to the pressures of competitive sports, something changed in the year leading up to the Games—he started *really* winning. First quietly, then loudly, commentators called him a favorite for gold, with two-time Olympian Johnny Weier dubbing him "America's hope." For Nathan, this brought a "new kind of pressure." Soon, big-name sponsors like Coca-Cola and Nike signed him, Vera Wang was designing his costumes, and the media hounded him for interviews.

The day Nathan secured a spot on the Olympic team, he was at San Jose Arena with his elated sisters. When they asked how he was feeling, he tried to match their enthusiasm. But instead of excitement at achieving his lifelong dream, he felt crushing waves of dread. Under the weight of the world's expectations, the expectation to win gold was powerfully undermining his confidence.

In interviews, Nathan made sure to give the "right" answers, like how happy he was to realize his dream and how honored he was to be competing at the Games. But with each passing day, he became more anxious, having internalized the pressure from the media, his corporate sponsors, and his coaching team. *What if I can't make it happen?* Four brilliant lines from one of my favorite musicals, *Hadestown*, beautifully explain this experience:

The dog you really got to dread
is the one that howls inside your head
it's him whose howling drives men mad
and a mind to its undoing.

Before the Olympics, Nathan's career had been overwhelmingly intrinsically motivated. Of course, he always wanted to win, but he had generally focused on performing at his best. Now with this new breed of self-doubt howling in his head, he was seduced by the extrinsic goal of proving his competence. He began to believe that he "absolutely, positively had to win," because "if I couldn't win Olympic gold, what was I worth?" (He's not alone. Research shows that a surprising number of elite athletes are literally willing to die for an Olympic gold medal!) The moment Nathan adopted this "gold or bust" goal, he was no longer skating for himself—he was skating to prove himself. Pizza had won over salad, and shadow goals swallowed three-to-thrive needs.

The month before the Games, Nathan's stress was at an all-time high. The morning of the team competition, he was a nervous wreck. Shaking so violently that he barely heard his name called, he skated out onto the ice, glanced at the Olympic rings, and froze.

We know what happened next.

After Nathan's dismal finish, he ran downstairs to the practice rink, still in his costume, and began running through all the jumps he missed. When another skater asked, "Didn't you just compete?" Nathan willed him not to see the tears collecting in his eyes.

Then, days later, in the throes of his disappointment and grief, something unexpected happened. The night before his final event, the free skate, Nathan came to the heartbreaking realization that Olympic gold was a mathematical impossibility. Once his shadow goal went up in smoke, only one thing remained: the "love of the game" that fueled his success in the first place. Tomorrow, he decided, he would go out there *on his terms*—and he would skate his heart out until the music stopped.

On the day of his free skate, Nathan sailed out onto the ice and became the first skater to perform five perfect quadruple jumps in an

Step 3: Spot Your Shadows

Olympic competition. Though he didn't medal, his record-breaking performance marked the beginning of a four-year process of inoculating himself against external pressures and finding a more stable measure of self-worth that didn't depend on winning. The strategy paid off. Four years later, when he took the ice at the Capitol Indoor Stadium in Beijing, China, he wasn't focused on medaling but on his love of the game. Nathan Chen walked away from the 2022 Olympic Games with not one but *two* gold medals, becoming the first American figure skater to do so since 1952.

SPOTTING SHADOW HABITS

A few years ago, I met my own shadow in the middle of evening rush hour. It was pitch-black at 5:30 p.m. as I, driving home from a doctor's appointment, sped down a busy four-lane artery into downtown Denver. Without warning, the car in front of me screeched to a halt. After a few confusing seconds, it finally swerved into the next lane and sped off.

That's when I saw it: the reason the vehicle had stopped, staring wide-eyed at my headlights. In the middle of the road was the biggest rabbit I had ever seen. I slammed on my brakes.

Maybe it's because nearly all small animals remind me of my beloved six-pound rescue poodle, Fred, but the terror in the old rabbit's eyes broke my heart. Turning on my hazards, I began trying to chase her out of harm's way. I flashed my lights, honked my horn, opened my window, and leaned out. Flailing my arms, I feebly shouted, "Get out of the road!" But there she stayed, staring into my headlights, stuck in faint-or-freeze mode.

What should I do? If I drove away, I'd be a party to bunny slaughter. But the two lanes on either side of us were still moving and it wasn't safe to get out and help her.

My mind suddenly flashed back to that rainy afternoon when, in the depths of my illness, I forced myself to confront my emotional pain and wrote "helpless" on that Post-it. At that moment, shouting at the rabbit in the road, there was no better word to describe how I felt. Looking at the poor creature, rigid with fear on the pavement, I had the urge to give up and leave her to her fate.

Wait. No. This time, I would make a new choice. *I was going to save this rabbit.*

My heart beating out of my chest, I slowly exited the car, signaling to the two vehicles approaching me on either side to stop. Amid the chorus of honks, I gestured wildly at the rabbit to make sure they saw her. The driver to my left rolled his eyes and drove off, but the driver on my right stopped. I rushed up to the rabbit and gently nudged my foot against her. "You need to go. *Now.*"

Slowly at first, then quickly, she started hopping toward the right lane, and then out of the road, hurling herself up to the grassy parkway. Breathing a sigh of relief, I returned to my car, feeling equal parts terror and triumph. While a little worse for the wear, the rabbit and I had both survived!

When I got home, I called Sarah, my best friend of three decades. When I told her what I'd done, she had trouble hiding her surprise. "The Tasha *I* know would've driven away and regretted it for the rest of her life."

Hmm. That was interesting.

In the following days, I pondered what Sarah said. I'd always seen myself as a take-charge person who *absolutely* would have rescued the rabbit. But Sarah's comment made me realize that of late, perhaps something had changed. Sensing I was on the brink of some Big Realization, but unsure how to proceed, I took a page from Carl Jung, opening yet another Word document and brainstorming several questions to help me delve into the depths of my shadows.

Usually, the most tangible representation of our shadow goals are **shadow habits,** which are how shadow goals want us to act. Here, I found one question to be particularly useful: **How is my current behavior different from when I'm at my best?** That's when the significance of Sarah's comment hit me. At some point in the months since the onset of my mysterious illness, I had developed a habit of giving in to my feelings of helplessness, only to regret it later.

Step 3: Spot Your Shadows

But how could this be? Making a living wrestling the strong personalities of CEOs, I hadn't earned my favorite nickname, "The Velvet Hammer," by giving in when the going got tough. Lately, though, I realized that the best self I brought to my work had stopped showing up outside of it. As I interrogated this insight, my endless doctor's appointments replayed in my mind, each one like a broken record reinforcing my helplessness. I'd start by summarizing my symptoms; then, a highly credentialed specialist would respond with a furrowed brow and ominous "Iiiinteresting . . ."[4] A parade of pricey tests would all come back normal. The specialist would declare me "perfectly healthy" and "completely fine." Then, they'd insist I was far too young to *actually* be sick, suggesting I needed to "relax," "meditate more," or (my favorite) "maybe stop being so dramatic."[5] Their thinly veiled message? "If I, an esteemed medical expert, can't find anything wrong with you, there *is* nothing wrong with you."

While I'd broken with type to save the rabbit, a clear shadow habit was showing up in my illness journey: to cope with the lack of agency I felt, I had (unbeknownst to my conscious mind) defaulted to textbook prevention behaviors. Namely, I was giving up on exercising the agency I *did* have by getting on the most comfortable path and staying there, waiting for some mysterious external force to swoop in and change the game.

At that moment, I realized something that almost knocked me over. *There was no cavalry coming to save me.* If I continued this way, things would never improve. While the epiphany stung at first, it was oddly invigorating. I now knew what Carl Jung meant when he said that gaining awareness of our shadows helps us assimilate them into our self-concept without judgment, making the space for self-acceptance and reinvention. By coming to terms with my most self-limiting shadows, I could finally feel myself waking up to a better way.

---

4. In a healthcare setting, you never, and I mean *never*, want to be the "interesting" patient.

5. True story: Until my diagnosis, I spent more than thirty years believing that everyone else was *also* in ten-out-of-ten pain every day, but I just wasn't as good at pushing through it as they were.

CHAPTER 7: KEY TAKEAWAYS AND TOOLS

"When we become aware of our shadows, we pave the way for self-acceptance and reinvention."

- **Shadows**: Our instinctive responses to persistently thwarted needs that turn us into the worst version of ourselves.
- **Shadow goals**: "Quick fixes" for self-worth, power, or approval that actually make need thwarting worse (see the following page for examples).
  - **Compensatory motives**: "If I can't feel confident / choiceful / connected, at least I can . . ."
    - **Protect**: Looking for antidotes to guilt, shame, and bruised ego.
    - **Prove**: Seek evidence to prove that I'm worthy, powerful, and loved.
    - **Prevent**: Trying to avoid feeling bad by escaping, ignoring, or downplaying.
  - **Extrinsic motivation**: Acting based on pressure, guilt, or rewards, which thwarts needs.
  - **Intrinsic motivation**: Acting from authentic choice, which deepens need fulfillment.
- **Shadow habits**: The way shadow goals want us to act.
  - **Shadow habit-seeking question**: "How is my current behavior different from when I'm at my best?"

Sample Shadow Goals

Compensatory Motive	Thwarted Confidence	Thwarted Choice	Thwarted Connection
PROTECT: Fighting back to avoid guilt, shame, or blame.	• **Defensiveness:** A drive to devalue others' perspectives.	• **Rebellion:** A drive to blindly defy expectations or rules.	• **Spite:** A drive to hurt others who've hurt us. • **Aggression:** A drive to lash out.
PROVE: Seeking external evidence about ourselves or our place in the world.	• **Achievement:** A drive to achieve greatness at all costs. • **Rewards:** A drive to amass material rewards like wealth or image. • **Perfection:** A drive to meet excessively high standards.	• **Dominance:** A drive to exercise control over others. • **Restriction:** A drive to control or limit one's own choices.	• **Popularity:** A drive to be liked and socially powerful. • **Validation:** A drive to be accepted by others.
PREVENT: Denying experiences or devaluing our needs to feel better right now.	• **Escape:** A drive to avoid situations that hurt our confidence. • **Inertia:** A drive to maintain the status quo. • **Ignorance:** A drive to ignore difficult (self)-truths.	• **Harmonizing:** A drive to do what we "ought" to make things "easier." • **Giving Up:** A drive to surrender without resistance.	• **Pretending:** A drive to avoid rocking the boat by wearing a social mask. • **Seclusion:** A drive to close ourselves off from others.

CHAPTER 8

Step 4: Pick Your Pivots

> In the depth of winter, I finally learned that
> within me there lay an invincible summer.
> —ALBERT CAMUS

At age thirty-six, Isabel married a worldly and accomplished diplomat. She knew deep down that he wasn't the life partner her younger self would have chosen, but with a dwindling number of childbearing years ahead of her, she didn't want to be picky. And her new husband did tick some boxes: he seemed to be well-read, confident, interesting, funny, and shared her love of literature and music. Plus, with a career that took him to far-flung locations, Isabel would get to travel the world by his side. Certainly, she figured, his cold and aloof demeanor wasn't anything that a little TLC couldn't fix.

Envisioning a future brimming with freedom and choices, she set aside her thriving career to embrace the nomadic life of a diplomat's spouse and later the task of raising their three sons. But as the years marched on, the idyllic life Isabel imagined faded out of view. Her husband's affection, she discovered, came at a price: prioritizing his needs over hers and faithfully obeying his rules. Over time, balancing these demands became increasingly challenging, especially as her boys entered their tumultuous teenage years. But Isabel soldiered on, determined to avoid voicing any dissatisfaction or disagreement that was sure to set him off.

That is, until one day, when an unexpected announcement shattered the family's delicate balance. Her husband came home from work and casually announced they'd be leaving everything behind—again—for a second lengthy posting in Africa. There was no discussion, no opportunity for input, and no consideration for how such a move would impact Isabel and their kids. The matter was already settled.

For Isabel, this was a bridge too far, and not just geographically. It was her actual breaking point. So, she did something she'd never done in two decades of marriage.

She said no.

When we find ourselves caught up in shadows and unwittingly contributing to our own need frustration, it's officially time to pivot. **Pivoting means proactively moving away from old, familiar shadows and building new paths to need fulfillment.** Whether these shifts happen in bold, life-changing flashes (like breaking free from a toxic relationship, as Isabel did) or more quietly or gradually (like reclaiming our leisure time or reshaping daily routines), they are generally inspired by a sentinel event that galvanizes us to go "all in" on a new shatterproof goal.

SENTINEL EVENTS

Imagine going to the hospital and your surgeon operating on the wrong part of your body, or a nurse giving you medication to which you are allergic. In healthcare, these are examples of "sentinel events," or situations that create unexpected patient harm, prompting the need to immediately investigate and course correct. Similarly, in our journey to become shatterproof, **sentinel events** are unmistakable warnings that force us to confront the true toll of our shadows, prompting a shift in strategy to prevent anything similar from happening in the future. Throughout this book, we've seen several examples of sentinel events: from Jan's epiphany on the stairs, to Carl Jung's decision to learn from his inner torment, to Nathan Chen's commitment to free-skate freedom.

I recently turned the Shatterproof Road Map into an exercise for an executive team. Amid a multiyear strategy and culture transformation project, I asked each executive to share a story where they found an unexpected opportunity in a crisis. As we tried to figure out what everyone's stories had in common, the COO—a thoughtful man of few words—eloquently summed it up: "**A singular moment of clarity where we chose to become an active participant in our own lives.**" His words captured the final step of the Shatterproof Road Map so brilliantly that they still give me chills.

And while I developed this concept as a researcher and road-tested it as a consultant, I only discovered its true meaning once I began walking this path myself. My own sentinel event struck the moment I finally realized the dangerous holding pattern I was in: endlessly unhelpful doctor's appointments, doing what I was told, and clinging to the belief that only my specialists could save me. As "safe" as that felt at the time, I could now see that it wasn't protecting me; it was imprisoning me.[1] This made my choice crystal clear; I could either keep being a passive bystander—and possibly die—or I could chart a new course and become the CEO of my medical journey.

SHATTERPROOF GOALS:
NOVEL PATHS TO NEED FULFILLMENT

Sentinel events reinforce which three-to-thrive need—confidence, choice, or connection—currently feels most limited in our life. The next step is to ask: What must happen for that need to feel more fully met?

When I pose this question to clients and keynote audiences, responses often revolve around the desire for more recognition, flexibility, and appreciation at work, less pressure, mustivation, and conflict in their personal life, or some combination of the two. Whether we yearn for a more empathetic partner, a less demanding boss, or a more fulfilling career, we commonly assume that changes to our environment are the key to need fulfillment. This, too, has been self-determination theory's prevailing assumption for almost five decades. Researchers

---

[1]. To quote the great Canadian rock philosophers, Rush, when we face the choice between "phantom fears and kindness that can kill," in choosing not to decide, we "still have made a choice."

Step 4: Pick Your Pivots

traditionally emphasize the role of external factors like parenting styles, teacher behavior, and workplace dynamics in supporting or hindering human needs.

That is, until 2019, when one groundbreaking idea emerged. In an otherwise unassuming presentation at the seventh annual Self-Determination Conference in Amsterdam, then–graduate student Nele Laporte and her team proposed that humans aren't merely "passive recipients" of their environment. Instead, they explained, we can actively shape our own needs *regardless of our external circumstances*.

This is the insight at the heart of a wonderfully straightforward process called **need crafting,** which was developed by Laporte and her colleagues. First, we identify our unmet need(s) and the obstacles to getting them met (as we did in Chapter 6). Then, we pivot by choosing new goals and habits to maximize satisfaction of that need.

Early need crafting research crackles with promise. In Laporte's study of over eight hundred adolescents, those who consistently practiced need crafting reported higher confidence, choice, and connection need satisfaction, less stress and depression, more zest for life, fulfillment, and better moods. Perhaps the study's most striking finding was that need crafting was a slightly *stronger* predictor of need satisfaction than environmental factors like supportive parenting. Considering the nearly half century spent examining the central role of our environment in need satisfaction and frustration, this finding is intriguing indeed.

Additional research shows that crafting our needs boosts mental health, not just in everyday situations, but also during challenge and crisis. For instance, a study during COVID lockdowns found that a ten-day online need-crafting program helped participants feel more satisfied, energetic, and motivated while reducing their stress. Other work confirms the power of need fulfillment in tough situations like financial hardship, dangerous living conditions, and even displacement due to war. The takeaway? **We possess the power to transcend the limitations of our environment by proactively shaping our own needs.**

Our study's early interviews on how people handled "bad things" didn't specifically look at need crafting, but the concept popped up in a rather interesting way. Across multiple samples and situations, *only three responses* consistently predicted people growing forward through tough times—what we might call shatterproof outcomes:

- I made myself or the situation better.[2]
- I introspected to understand myself and make more authentic choices.
- I used the situation to make things better for others or the world.

What made these responses different from those that didn't predict shatterproof outcomes? For months, I was stumped. Then I noticed a pattern: make yourself better, make authentic choices, make things better for others. These closely mirrored the needs of confidence, choice, and connection. And *that's* when it clicked. From there, all I had to do was hit the research to find as many scientifically supported approaches to crafting our three-to-thrive needs as possible.

For instance, if you keep getting passed over for a promotion, you can focus on boosting your confidence by strengthening a certain skill or teaching someone something you're good at. If your boss micromanages your every move, you can deepen your choice through a hobby outside work that feels interesting, meaningful, or important to you. If you've just had a huge fight with a friend, you might focus on repairing the relationship or reinvest in another one you've been neglecting.

The following sections detail fourteen goals, grouped under six overall focus areas, scientifically shown to support our three-to-thrive needs. I call this model the **Shatterproof Six**. Review it, reflect on it, and choose the goal that feels right to you. (The end of this chapter also includes a list of sample actions to craft each three-to-thrive need.)

---

2. This included the goal of learning from the situation.

Step 4: Pick Your Pivots

THE SHATTERPROOF SIX[3]

✓ = primary need fulfilled; + = secondary need(s) fulfilled

Aim	Goal	Definition	Confidence	Choice	Connection
1. RISE: *Make myself better*	Mastery	Focusing on learning or refining a specific skill (process > outcome).	✓	+	
	Self-development	Committing to personal growth and expanding my horizons.	✓	+	
2. FLOURISH: *Make my life better*	Joy	Rediscovering my "love of the game" and immersing myself in things I like to do.	+	✓	
	Health	Maximizing my mental and physical health.	+	✓	
	Purpose	Doing things that bring me meaning and fulfillment.	+	✓	+
3. ACTIVATE: *Make things happen*	Agency	Making my own choices and being my own person.	+	✓	+
	Advocacy	Speaking up for myself and making my needs known.	+	✓	+
4. ALIGN: *Make authentic choices*	Self-awareness	Understanding and accepting who I am and how I fit into the world.	+	✓	+
	Authenticity	Expressing my values and showing up as who I really am.	+	✓	
5. RELATE: *Make meaningful connections*	Belonging	Building positive social bonds by meeting and affiliating with other people.	+	+	✓
	Closeness	Deepening close relationships by giving and getting support.		+	✓
	Forgiveness	Letting go of old grudges for my own well-being.		+	✓
	Spirituality	Connecting to something greater than myself.		+	✓
6. CONTRIBUTE: *Make the world better*	Service	Improving the world through generosity, contributing to the greater good, or making positive change.	✓ Deeply satisfies all needs		

---

3. Empirical evidence supporting each shatterproof goal as it related to need satisfaction can be found at www.shatterproof-book.com/goals.

After reviewing these options, the perfect goal might jump out immediately. Other times, a few follow-up questions may be necessary. First, consider your most frustrated need (the one you identified in Chapter 6) and ask yourself: Which goal would put me on the most authentic path to restoring this need? (Notably, most shatterproof goals address multiple needs, but I suggest focusing on your biggest thwarted need.) If you're still torn, consider the shadow goal(s) you isolated in Chapter 7 and ask: Which shatterproof goal is the most natural foil to my shadow goal(s)?

Research shows that there are many "right ways" to achieve shatterproof outcomes, but when selecting a goal to get you there, a few caveats are in order. First, for the biggest payoff, make sure to choose intrinsic goals that are personally important, enjoyable, or challenging (in a good way).[4] After all, how can we successfully reinvent ourselves if we don't authentically buy into what we're trying to change? (Or, as my dear friend and author of *Wonderhell*, Laura Gassner Otting, preaches, we can't be hungry for someone else's goals.)

Second, to minimize the chance of becoming frustrated or discouraged, try to pick the shatterproof goal that feels as realistic as it does motivating. And above all, your goal should be sustainable over time. One study revealed that shatterproof goals enhanced well-being only when participants remained committed to them over a six-month period. Shatterproof goal-setting therefore can't be a "one-and-done" exercise and ongoing benefits require ongoing commitment. This is why selecting our shatterproof goal is the pivot—not the end, but a new beginning.

HOW STRATEGIC EXPERIMENTS TEST SHATTERPROOF HABITS

After Isabel decided to pivot by openly defying her controlling spouse for the first time, the consequences were swift and brutal. He stopped

---

4. Helpfully, the mere act of freely choosing something—a goal or anything else—nurtures both confidence and choice. In one study, participants given autonomy to choose between two sets of almanac questions felt more competent before starting the task than those who weren't given a choice (Rebecca A. Henry, "The effects of choice and incentives on the overestimation of future performance." *Organizational Behavior and Human Decision Processes* 57.2 [1994]: 210–25).

Step 4: Pick Your Pivots

speaking to her and divorce proceedings were promptly initiated. Determined to exact punishment for her audacious act of refusal, he brandished the facade of doting husband and courtroom victim. In the end, Isabel walked away with very little, and the man who'd vowed to cherish and protect her twenty-three years earlier walked away with their life savings and never looked back.

Isabel was forced to move more than one thousand miles away from her children. Hardly able to pay her bills, she'd lay awake at night, staring at the ceiling while gripped by a paralyzing fear of becoming homeless. The few fitful hours she managed to sleep brought recurring nightmares of her ex-husband strangling her, burning her clothes, or hurling a hand grenade into her bedroom. (One can only imagine what Carl Jung would have made of those midnight blockbusters!) She had no words to describe the pain of having the person at the center of her life—the man she supported and trusted and the father of her babies—determined to destroy her future.[5] The good news was that Isabel was no longer suppressing her emotions; the bad news was that she had become paralyzed by them.

She reached a turning point when an old friend wondered aloud whether Isabel's agony was becoming a sort of addiction. In a flash of clarity, she realized her friend was right. She'd been so overcome by the betrayal that she was turning into a sour person, infected by hatred and pain. After so many years of loving a narcissist, she'd lost touch with the kind, competent, and loving person she was.

In this sentinel moment, Isabel chose the shatterproof goal of agency. If she was to leave her shadows behind, it was time to stop existing in survival mode and start taking steps to build a **new, good life**—perhaps not the life she had expected, but one that would create a parallel path while prioritizing *her* needs and well-being.

---

5. There's a growing area of psychology focused on "relational trauma," or "betrayal trauma," which occurs when the person we trust most betrays or hurts us. It's easy to underestimate the toll of such trauma. I recently spoke with a friend who just found out that her kind, supportive husband of fifteen years had been leading a secret life and cheating on her for their entire marriage with strangers and sex workers. Years earlier, she'd been diagnosed with PTSD after a long tour in the Gulf—she compared her husband's betrayal to "one hundred Iraqs."

Despite being invigorated, Isabel was unsure how to achieve this goal. As a jumping-off point, she asked, *What would have to happen for me to truly feel like an active agent of my own life?* Without hesitation, she jotted down "connection with my children and loved ones, the capacity to be financially self-sufficient, and the strength to prioritize my self-belief and well-being." With these answers in hand, she challenged herself to craft three shatterproof habits:

1. Remain a daily source of love for my fantastic boys and amazing friends.
2. Earn $5,000 a month to support my family.[6]
3. Wake up every morning with a big smile that celebrates my freedom from a toxic relationship.

These habits, the very definition of Isabel's core values, offered her new energy and focus. Now it was time to road test them. Enter strategic experiments, the iterative process of making new, shatterproof habits a long-term part of our lives. **If shatterproof habits are the seeds of our shatterproof goals, strategic experiments are the nurturing sunlight that helps them take root and start to sprout.**

Instead of grand gestures, Isabel focused on taking purposeful, incremental steps each day to prioritize her family, financial stability, and overall well-being. Eventually, she found a job at a publishing company, and while cash was still very tight, the confidence and sense of agency she derived from her work was worth more than anything money could buy. That was when everything around her started to change.

She concentrated on spending as much time as possible with her children, and planning inexpensive holidays together whenever she could

---

6. While a monetary goal might appear to be an extrinsic, or shadow goal, the motive behind resource-related goals can sometimes swing them into intrinsic (aka shatterproof) territory. Here, Isabel was focused on earning a living so she could make her own authentic choices and support her boys rather than on making $5,000 to prove that she was "successful." (See Athanasios Mouratidis et al., "Beyond Positive and Negative Affect: Achievement Goals and Discrete Emotions in the Elementary Physical Education Classroom," *Psychology of Sport and Exercise* 10, no. 3 [2009]: 336–43.)

Step 4: Pick Your Pivots

afford it. After all, "you don't need riches to enjoy a lively conversation in a café with a friend or share a home-cooked meal with your children," she wisely told me. She looked after the friends and family members who'd always been there for her and tried to forget those who hadn't. And most of all, she found a way to acknowledge past mistakes while also forgiving herself for the poor choices she'd made in her former life.

One year later, Isabel stood triumphant, having proven that she could become a reliable source of love and support for herself and the people she cared about, completely on her own. This love nurtured her soul in ways she could have never imagined. Isabel was astonished that she hadn't just rebuilt her life; she'd crafted an even better one than before.

WHEN ONE SHATTERPROOF HABIT IS *PLENTY*

Sometimes, like Isabel, we can identify several shatterproof habits to experiment with. Other times, the victory lies in simply choosing one. (Remember, it's not a crime to make your development easy.) In my own case, I was so overwhelmed with the enormity of my illness—and the task of saving my own life—that I decided one shatterproof habit would have to be enough. But which one?

Perhaps I could launch an outright fight: submit formal complaints against dismissive doctors, try to get my story covered on a morning talk show, cause a fuss on social media, or even start a political action group. I knew there were people out there who'd been ignored, patronized, and dismissed by the traditional medical establishment, just as I had been. Or, what about becoming my own doctor? I could put my career to the side for a few years and go to medical school!

It was at this point that I knew it was time for a hot mug of calm the hell down, Tasha. This was obviously not realistic.

Or was it?

Maybe a more modest version was possible. I couldn't quit my job to become a doctor, but I could invest thirty minutes a day in self-directed medical education. Maybe I could draw on some of the skills I'd honed in graduate school: my tenacity, my passion for "hitting the literature," my ability to synthesize complex information. And even if thirty minutes a day wouldn't lend me the level of expertise I'd need to conclusively diagnose and treat myself, it could help me formulate evidence-based theories to present to my doctors.[7]

So that's what I did. For half an hour every evening—sometimes more, but never a minute less—I hit Google Scholar and added to my growing pile of printed medical research. A month or so later, I was pleasantly surprised to be making real progress. And relatively speaking, I was actually feeling kind of great. While the physical illness was as gruelingly present as always, for the first time in forever, I felt like *me*. The simple pivot from helplessness to power was filling me with purpose, energy, and pride.

I compiled a list of ten rare diseases that most matched my symptoms and started to research the information I needed to either diagnose or rule each one out. Like a good detective, I decided it would be important to eliminate as many possible suspects as I could, starting with the most obvious culprit: autoimmune disease. For this, I'd need a rheumatologist. After somehow managing to talk my way onto the schedule of one of the best in Colorado, I arrived at my appointment with a short, carefully written report. In it was a symptom summary, a personal and family medical briefing, and a few pertinent appendixes (e.g., "Tasha's Bones: A Brief History"). My goal was to communicate that (a) I knew what I was talking about, and (b) I refused to be given the "big shush."

I handed my report to the receptionist and flinched inwardly at her expression. I couldn't quite tell. Was that contempt I saw flashing across her face? Was it a panicked moment of "Oh, dear, we have a crazy woman in the building?" Or was it simple, innocent surprise?

As I sat in the waiting room, my heart pounding, something felt off. What if my report made matters worse, landing me the label of "difficult"

---

7. The US healthcare system is full of wonderful, skilled, heroic professionals who are facing a system that pressures them to focus on patient *volume* over patient *helping*. I know this not only as a patient, but also because I spent three years heading up leadership development in a top Denver hospital before I started my business.

Step 4: Pick Your Pivots

patient in addition to being an "interesting" one? I tried a new mantra. Breathe in: *stick to your guns*. Breathe out: *stick to your guns*.

Fifteen minutes later, I sat on the exam table in a large, windowless room papered with posters of autoimmune diseases. When Dr. Goldstein entered holding my report, I could see some handwritten notes scribbled in the margins. I felt like a student who was about to receive a stern talking-to from the principal. My mouth felt incredibly dry.

"Hi, Tasha," she said. "So I took a look at your document." She glanced briefly down at it. "It sounds like you've been through a lot. How would you like to start?" In my decades as a patient, I couldn't remember anyone asking me this question.

I nervously cleared my throat. "I have two goals," I said. "The first is ruling out an autoimmune disease."

"Okay."

This was my moment. I presented my theory of the case, citing peer-reviewed studies and complex medical terminology that only a rheumatologist would understand.

"That makes sense," she said, tentatively. She paused for a moment, in thought. "Yes, that makes a lot of sense."

I almost fell out of my seat.

After conducting a physical exam and reviewing my labs, Dr. Goldstein announced, "You said ruling out an autoimmune disorder is your goal one. Well, from what you've told me, and what I'm seeing, it's highly unlikely you have one. You said you had a second goal?"

I explained that I believed I had a genetic connective tissue disorder that affects the functioning of the body's two proteins, collagen and elastin. Because these proteins are found in every system of the body, the disease can trigger a raft of seemingly unrelated issues that don't "fit together." As far as I could tell, my symptoms fit best with a disorder called Ehlers-Danlos Syndrome (EDS).

At this point, I paused to make sure she wasn't about to escort me out of the building. One thing I knew about EDS was that it is incredibly controversial. Most doctors are (incorrectly) taught that it's so rare, they'll never see a single case in their career.[8] Since it doesn't present with obvious signs in labs or imaging, along with its numerous and varied symptoms, people with EDS are often questioned, challenged, or simply not believed. My whole life, I'd endured this skepticism firsthand as doctors and loved ones were quick to label me dramatic or high maintenance, and slow to believe that what I was experiencing was, in fact, real. At times, it left me wondering whether my symptoms really *were* in my head. And if the struggle to be believed is hard, getting an actual EDS diagnosis is even harder, commonly taking ten to twenty horrifying years.

Mercifully, though, Dr. Goldstein didn't flinch. Instead, she gestured, go on.

There are thirteen kinds of EDS, I continued, twelve of which had genetic markers. I believed I had the thirteenth, something called hypermobile EDS. I listed the telltale signs: unusually flexible joints; debilitating muscle, joint, and nerve pain; degenerative changes across multiple systems of my body.[9] From having conducted a thorough family medical history, I knew that both my mother and grandmother suffered the same seemingly unrelated symptoms (though my mother's, thankfully, are less severe).

Hypermobile EDS also has two primary cotravelers—symptoms I'd always had, but were now worsening to the point of unmanageability.[10] The first condition, postural orthostatic tachycardia syndrome, disrupts the body's automatic functions like heart rate, breathing, temperature, digestion, balance, memory, and the incredibly helpful function of *sending oxygen to the brain*. With the second, mast cell activation syndrome,

---

8. Recent research estimates an incidence of *at least* one in five thousand people (Germaine L. Defendi, "Genetics of Ehlers-Danlos Syndrome," Medscape Reference, updated March 14, 2024, https://emedicine.medscape.com/article/943567).

9. According to EDS expert Dr. Predeep Chopra, "Patients with EDS present with one of the most chronically painful conditions in medicine" (Anne M. Maitland, *Disjointed: Navigating the Diagnosis and Management of Hypermobile Ehlers-Danlos Syndrome and Hypermobility Spectrum Disorders* [San Francisco: Hidden Stripes, 2020]).

10. These three syndromes co-occur so often that the EDS community calls them "the Trifecta." Some doctors believe they're all part of the same illness.

the body's mast cells[11] chronically overreact to everything from foods to smells to stress, causing widespread inflammation as well as neurological, vascular, endocrine, skin, and intestinal problems.

When I finished, Dr. Goldstein was silent. Then she stood up, walked over to me, and said softly, "I believe you."

I blinked back tears.

"EDS is a complicated disease," she said. "And I can tell you that you already know about one hundred times more about it than I do." I beamed. "I can't diagnose EDS. But based on your clinical exam, I can diagnose you with joint hypermobility syndrome, which is a kind of umbrella for diseases including EDS."

"I'll take it!" I replied, perhaps a little too enthusiastically.

Wrapping up the appointment, she asked, "Do you have any questions?"

I racked my brain. What would someone radically in charge of her health ask right now? "Well, I'd appreciate a concrete next step. Since you don't diagnose or treat connective tissue diseases, who do you know who does?" She offered the name of a specialist, Dr. Jill Schofield—a global expert in EDS and its related conditions whose office was somehow just twenty minutes from my home—along with the warning that it could take years to get in.

Driving home, I felt great. Okay, so I didn't have the definitive answer I'd hoped for. But through my first strategic experiment, I'd stumbled on a new way to work with doctors *and* lucked out with one who was willing to hear me. From then on, my printed medical summaries became a staple of all my doctor's appointments, whether it annoyed the doctor or not. This eventually snowballed into four large, dog-eared, color-coded binders containing all of my summaries, medical records, and relevant research on each of my conditions. (Yes, dear reader, this doctor became *that patient*.)

A few weeks later, a miracle occurred. After being disheartened that Dr. Schofield didn't have an available appointment for fifteen months, a

---

11. Mast cells are white blood cells present in the body's connective tissues and mucous membranes that identify and engage with threats, like viruses. The systems in the body with the most mast cells include the skin, lungs, blood vessels, nerves, and intestinal tract.

very generous friend managed to get me into a world-renowned facility with an EDS clinic—and my appointment was scheduled for the very next week.

Four days later, as I strode through the center's automatic doors, I couldn't have felt more different from the broken person who'd apologetically shuffled in to see Dr. Goldstein. This time, I knew exactly what to do. And between the center's expertise and my careful preparation, I was sure I would finally get my diagnosis.

CHAPTER 8: KEY TAKEAWAYS AND TOOLS

"We have the power to transcend the limitations of our environment by proactively shaping our own needs."

- **Pivoting**: Proactively moving away from familiar shadows and toward new paths to need fulfillment.
  - **A new, good life**: Pivoting away from the life we planned and creating a new path where we prioritize our needs, our well-being, and find new happiness.
- **Shatterproof goals and the Shatterproof Six**: Scientifically supported objectives to craft our three-to-thrive needs like mastery, joy, purpose, authenticity, closeness, and service (see page 123 for a full list).
  - **Sentinel event**: A moment where we confront the true cost of our shadows and choose to become an active participant in our own lives.
  - **Need crafting**: Finding new ways to meet our own needs.
- **Shatterproof habits**: One or more regular behaviors that will help us achieve our shatterproof goal.
  - **Strategic experiments**: The iterative process of making new, shatterproof habits a long-term part of our lives.

What's next? The next section will be a deep dive into each of our three-to-thrive needs. But first, you might be interested in learning where you are with the practices of the Shatterproof

Step 4: Pick Your Pivots

Road Map. Check out Appendix D for a self-assessment. Also, to keep things practical, here are a few simple, scientifically supported actions you can take to start need crafting today:

Confidence	Choice	Connection
• Practice a simple activity that you're good at (crossword puzzles, running, singing, etc.). • Do something challenging that you'll feel effective in doing. • Teach someone something you're good at. • Do something that allows you to express yourself creatively.	• Make one small but meaningful choice today about how you'll spend your time. • Do something that will help you prioritize your needs. • Take a break from an activity that you do because you have to, but that you don't value. • Share how you *really* feel to someone important to you.	• Tell someone that you care about them. • Express gratitude to someone important in your life. • Help someone with something. • Do something fun with a loved one. • Do something that helps you feel connected with nature.

PART THREE

Shatterproof Transformations and Tools

Chapter 9

Crafting Confidence

> "Self-possession, that's what it was. This was someone who knew her own mind, her own worth."
> —Kate Morton, *The Clockmaker's Daughter*

When Stephanie Szostak left her native France for the United States, her plan wasn't to become a movie star. After earning her marketing degree, she landed a job at Chanel, which opened an unexpected door to modeling. This led her to enroll in an acting class, only to discover that pretending to be someone else made her feel more like herself than she could ever imagine.

But success took time. Stephanie paid her dues for nearly a decade, managing to land several smaller parts, including the iconic role of Jacqueline Follet in *The Devil Wears Prada*.[1] Then, after a last minute audition, several callbacks, and a nerve-racking chemistry read, Hollywood finally opened its doors. With a $69 million budget, *Dinner for Schmucks* starred comedy greats Paul Rudd and Steve Carrell—and Stephanie Szostak. On a warm fall Los Angeles morning, the actress drove her rented Ford Focus onto the Paramount lot to film her first leading role in a major studio movie.

It should have felt like the first day of the rest of her life as a movie star.

But as that first day unfolded, Stephanie found herself increasingly overwhelmed. Not because of issues with her director, co-stars, or script—

---

1. A movie for which I cannot possibly overstate my love.

which were all wonderful—but because in this sea of famous faces, she was an outsider. Though the spotlight was on her, she was by far the least experienced person on set, and the bullying voice in her head knew it. After each take, it berated her: *That sucked. They're gonna think it sucked. They're gonna think you suck. You're gonna get fired. And then your team is going to drop you. Because you suck, Stephanie Szostak, and everyone knows it.* Desperate attempts at silencing her inner voice only seemed to increase its volume. Stephanie felt like a balloon slowly floating away, disconnected from herself and everyone around her. How had her biggest break so quickly become her biggest source of stress?

In Chapter 6, we learned about our three-to-thrive need for confidence. At a fundamental level, **confidence has two elements: a sense that we are doing well and a sense that we are getting better.** When this need is fulfilled, we feel effective and capable. When it isn't, self-doubt can cast a dark shadow. We might therefore see the need for confidence as falling on this spectrum:

Self-Doubt  Competence

Full disclosure, Stephanie is my friend, so I may be biased, but her charming and funny performance in *Dinner for Schmucks* never even hinted at her inner turmoil.[2] And that's exactly the point: **our sense of confidence is surprisingly unrelated to our actual abilities.** As further evidence, I recently came across an article about my favorite vocalist, Josh Groban, the five-time Grammy Award–winning singer, songwriter, and Broadway superstar.[3] Earlier in his career, despite sold-out performances and several multiplatinum albums, the outwardly poised artist felt undeserving of his success. His inner monologue echoed Stephanie's: *You suck. What are you doing? Get your scrawny ass out of here.*

---

2. Side note: This experience became a turning point for Stephanie, leading her to seek out the skills and practices that ultimately helped her grow, which you can read about in her delightful workbook, *Self!sh*.

3. As my friends will tell you, I cannot say Josh Groban's name without adding "the voice of an angel!"

Crafting Confidence

Discrepancies between objective abilities and subjective perceptions are not unique to those who perform for a living; they've been found for athletic ability, academic performance, financial know-how, and more. And notably, it is our *subjective* confidence—not our objective performance—that most reliably predicts well-being.

In general, our self-perceptions can be out of sync with reality in two ways. We can overestimate our performance, or we can underestimate it. In normal times, the average person is more likely to be an overestimator (a finding I found so interesting that I wrote a book about it, *Insight*).[4] But when we're triggered, confidence can take a nosedive, causing us to seriously underestimate our true contributions.

While confidence triggers are different for everyone, we can identify several common culprits. Josh Groban and Nathan Chen (whom we read about in Chapter 7) were triggered by **expectations** to perform well, a trigger amplified by their increased visibility. At the other end of that spectrum, **monotony** can also be a trigger.[5] When we're not feeling challenged, we can't achieve our potential, which can lead to worries that we're not as effective or competent as we thought. Another trigger is **chaos,** including unclear expectations, unexpected shifts in our roles or responsibilities, and a general lack of consistency and predictability. Then of course, there are signals that our performance isn't up to par, like **setbacks, criticism,** and **inferiority,** which Stephanie experienced on set. (Flip back to page 103 for more on common confidence triggers.)

In sum, confidence triggers make us feel inadequate, inferior, demotivated, or like we're a failure—and this can lead to shadow habits like excessive self-focus, unnecessary competitiveness, and paranoia. At work and school, these habits are associated with negative self-talk, exhaustion, and burnout. In relationships, they lead to greater unhappiness, decreased

---

4. As an important note, overestimating ourselves is not the same thing as fulfilling our confidence need. The former is ego driven, while the latter is a fundamental psychological nutriment that actually reduces our ego.

5. People display a natural inclination for tasks just above their current level of ability (Fred W. Danner and Edward Lonky, "A Cognitive-Developmental Approach to the Effects of Rewards on Intrinsic Motivation," *Child Development* 52, no. 3 [1981]: 1043–52).

energy, and poorer life satisfaction. (For a list of common shadow habits for confidence, see page 157.)

But now for some good news: Even if it doesn't feel this way at the time, getting our confidence need met isn't as hard as we might think. In fact, confidence-crafting opportunities are all around us, if we just know where to look.

GETTING IN THE CONFIDENCE ZONE

Confidence is a visceral experience rooted in purposeful engagement with the world. Imagine tackling a significant task—a work project, health or fitness challenge, community or family goal, or any pursuit you find invigorating and important. When you dive in, you enter a state of "flow," the kind of intense concentration where you lose track of time while relishing in the focused pursuit of your objective. You feel motivated, interested, and energized. When unexpected challenges arise, you stay focused and committed, and should you fall short, you can appreciate your efforts while gleaning valuable lessons for the future.

Sounds wonderful, doesn't it?

Starting in infancy, humans are wired to seek deliberate activity. This is largely because complex mental processes like problem-solving, exploration, and insight were vital for our ancestors' survival, shaping us into "occupational beings" with an innate drive to put our time to good use.[6] All we need, then, are a few practical tools to help us channel this instinct into proactive need crafting.

Taming the Doubt Dragon

Grace was unlike any leader I'd ever coached. On our very first call, the recently appointed CEO of a thriving tech company readily admitted to experiencing constant, crushing impostor syndrome. Grappling with the twin triggers of expectations and chaos, she felt as though she was

---

6. Why else would a long wait at the DMV be *that* soul crushing?

making everything up as she went, all while fearing that a single mistake would cost her the confidence of her board and employees.

But as I interviewed Grace's board, direct reports, and employees—even her husband and best friend—I quickly discovered that her perceptions were vastly different from *their* reality. In every previous coaching engagement of my career, my interviews would uncover at least three or four (sometimes many more) ways the client was getting in their own way, however unintentionally. But with Grace, and for the first time ever, I'd only managed to find one.

One foggy fall morning, I sat in a sparkling corner office awaiting her arrival. Admiring the view of the San Francisco Bay, I couldn't help but wonder what was about to transpire in what I knew would be a most unusual meeting. In front of me sat a twenty-seven-page narrative report detailing the results of my twenty-six interviews on how Grace was showing up as a CEO and a human. The findings were unmistakable: she was killing it. But would my new client believe me?

When Grace appeared, dressed in a dark suit with chic round glasses and flawlessly styled chestnut hair, she was holding a copy of her report, which I'd emailed the previous night. Sitting down, she uncapped her pen like a straight-A student eager to take notes.

"As you know," I began, "I help highly successful executives get even better. I've been working on your case for weeks. And I have never struggled so much to learn how a client was getting in their own way. But don't worry. I've found it."

I paused for dramatic effect.

"Your singular limitation? Right now, it's what's going on in your own head." Grace looked back at me, puzzled. "Because you are *so* diligent, *so* smart, *so* motivated, and you care so much about being successful, it's causing you to internalize and intensify your own stress."

I proceeded to walk Grace through her shadow habits, as described by those closest to her: sleepless nights where the wheels wouldn't stop turning, seeing minor setbacks as evidence of failure, and a stifling fixation on the 1 percent of people who were unhappy with her at any given moment.

Another casualty of Grace's lack of confidence was her unsustainable work volume. She rarely took time off, even during weekends and vacations—a habit that had finally caught up with her a few months earlier when she started experiencing chest pains. Emergency room doctors diagnosed an irregular heart rhythm, emphasizing the urgent need to minimize her stress.

Grace sighed heavily. "This is spot on. You know, before my promotion, I was mostly managing my stress. But not so much anymore."

I asked what was weighing on her. "I'm completely crippled by self-doubt," she replied. "Every day, I feel like I'm drowning and everyone is disappointed in me. In my quietest moments, I'm not even sure I deserve the role." Then, after a brief pause, "My new job appears to be colliding with my lifelong need to be perfect."

I was impressed that my client had zeroed in on her key issue so quickly: she was indeed exhibiting textbook perfectionist behaviors.

"And what does 'perfect' mean to you?" I asked.

"It means there's no room for error," she replied without a moment's hesitation.

"What else?"

"It means if I'm not flawless, I'm a failure."

I paused, letting the insight sink in. "But Grace, that sounds completely and utterly exhausting."

"Yeah! Apparently, it's killing me!"

It was clear that my client's current struggles were rooted in the same quality that had fueled her meteoric rise. A quarter century earlier, she'd been hired to answer phones, then was promoted to an assistant role, quickly becoming the fastest typist anyone had ever seen. Several promotions later, she became a manager, whereupon she decided to get her MBA and (surprise!) graduated first in her class. In the years that followed, she continued to rise through the ranks.

Then, just seven months before our meeting, Grace's boss—on the verge of retirement but not quite ready to go—had made the unconventional decision to appoint her as his co-CEO. According to their arrangement, Grace would take on nearly all the responsibilities of running the company, and his job would be to support her, largely but not exclusively behind the scenes.

Before that promotion, Grace told me, she'd mostly peacefully coexisted with the part of herself that needed everything to be perfect. But lately, she said, "I just don't know what's wrong with me. I can't manage it."

"Let me stop you there," I interrupted. "There is nothing wrong with you. Your response is perfectly understandable. You're dealing with more stress and self-doubt than ever—so you doubled down on a pattern that's largely served you in the past." Grace nodded.

Here's the thing about perfectionists: when everything is going according to their lofty plans, they're soaring high internally while projecting success externally. But throw in unexpected challenges, and these sky-high standards suddenly become profoundly limiting. Because perfectionists tend to fuse their self-worth with their achievements, even small setbacks set off a spiral of self-doubt. More reactive than the average person when things don't go as planned, they frequently view shortcomings as proof of failure and asking for help as evidence of incompetence. Unsurprisingly, these beliefs are a recipe for self-criticism, guilt, and shame; rumination about perceived failures and a fear of future ones; and even disrupted sleep.

Perfectionism was previously thought to affect mostly high achievers, but recent research uncovers a broader and more concerning trend. When economist Thomas Curran and psychologist Andrew Hill examined data from over forty thousand American, Canadian, and British college students between 1989 and 2017, they found a steady rise in perfectionism in this general population over time. The across-the-board uptick aligns with broader societal shifts. As the idea of meritocracy—that anyone can achieve anything through effort alone—continues to gain traction, it creates societal expectations that we should be striving for higher and higher levels of success.

Curran and Hill note that not only are young people increasingly pressured by the growing demands placed on them, many believe that they should also be even more demanding of themselves. In this sense, we might see perfectionism as a kind of coping mechanism to counter life's ever-increasing standards. (Appendix E will help you discover your own perfectionist tendencies.)

Perfectionism also makes us especially vulnerable to **impostor syndrome,** where we feel incompetent despite objective evidence that we're doing well. Impostor syndrome prevents us from internalizing success and drives us to work harder than necessary to avoid feeling inferior. In fact, one recent study suggests that it's actually impostor syndrome—not perfectionism in and of itself—that most undermines our well-being.

As a fellow recovering perfectionist, I knew I could help Grace tame her doubt dragon and that we could pivot by focusing on just a few new beliefs and behaviors. Our next step would be to replace her shadow goal of perfection with one that would boost her confidence and leave the self-criticism, worry, and workaholism behind. She wisely chose the shatterproof goal of self-development. "I want to use this process," she declared, "to develop myself as a CEO—and a human—by owning my confidence and well-being."

This sounded, well, *perfect* to me. We sealed the deal with a handshake.

But given her history of crippling self-doubt, would Grace be able to accept that she was already doing a far, far better job than she imagined?

Find a Reason to Be There

Grace and I spent the day poring over pages of positive feedback from my interviews. Her colleagues and loved ones described her as "exceptional," "fricking smart," "invaluable," "a delightful human being," and "the best boss I've ever had." They universally noted a rare blend of abilities: extraordinary intelligence, masterful relationship building, and courage under fire. A natural empath who intuitively connected with people, she had a preternatural ability for sharing credit and supporting others. And especially in tough moments, she was composed and candid.

Just imagine having a boss like Grace. Now, imagine that they had no clue how amazing they were. Inconceivable, right?[7] Indeed, as I

---

7. Make sure you read that line in the style of Wallace Shawn from *The Princess Bride*.

penned in Grace's report, "The hardest part of her journey may be hearing that she's already an extraordinary CEO—and believing it." And for our first strategic experiment, that's exactly what we spent the day focusing on. Hours later, after Grace absorbed her accolades, I finally asked, "How does all this feel to hear?"

"Well, I've heard some of this before, in bits and pieces, and far less specifically. I just never believed it. But I can't argue with your process, since you've talked to *literally everyone*. So I think it feels . . . good," she said, allowing herself a small smile.

Research suggests that our sense of confidence is strongly influenced by how we think *others* see us, a phenomenon psychologists call **metaperception**.[8] To better understand metaperception, consider the last time you gave an important presentation. In gauging how well you did, you might have tried to read your audience's mind by observing their verbal and nonverbal cues (Did they seem content or annoyed? Engaged or checked out?) or, even better, by asking for feedback.

Indeed, in feedback-rich environments, forming accurate metaperceptions is relatively straightforward. But most environments are far from rich in feedback. We're either damned with vague or generic praise—like hearing "good job" but not what exactly we did well—or no praise at all. **When we don't know where we stand, we can't know whether we're competent!** And especially when our confidence is threatened, we tend to believe that people view us more negatively than they typically do.[9] As we saw with Grace, the shadow habit of self-criticism only made things worse, priming her to assume that others were as critical of her as she was of herself. This is why Grace would later call our "feedback day" the most important step in her pivot from impostor CEO to confident CEO.

---

8. This is in part because of our need for connection. One way we feel like we belong is when others "see" that we're effective (Thomas Curran and Andrew P. Hill, "Perfectionism Is Increasing Over Time: A Meta-Analysis of Birth Cohort Differences from 1989 to 2016," *Psychological Bulletin* 145, no. 4 [2019]: 410).

9. This can especially be true when we believe powerful people like bosses, teachers, or coaches aren't happy with contributions (David A. Kenny and Bella M. DePaulo, "Do People Know How Others View Them? An Empirical and Theoretical Account," *Psychological Bulletin* 114, no. 1 [1993]: 145).

"The fundamental thing I lacked," she said, "was a belief that I could be number one. Hearing how much I'm needed and how effective my stakeholders see me—I didn't believe it before, but I believe it now."

Of course, when we solicit feedback, there's no guarantee it will always be positive. But paradoxically, well-delivered negative feedback can also be confidence boosting. When people are candid in their critiques, we can usually trust their positive feedback more. Plus, when we know what we need to improve, we're more in control and can—if we choose—decide to proactively tackle those issues.

But especially in confidence-threatening situations, there is extraordinary value in learning our defining strengths through the eyes of others. Rest assured, you don't need a coach to do this:[10] you can do it on your own, using a practical tool called the **reflected best self exercise (RBS)**. Having used the RBS for as long as I can remember, I've never had someone *not* uncover multiple "reasons to be there"—that is, to recognize their intrinsic value and see that they have earned their success. Research shows that the RBS boosts subjective confidence, renews psychological resources, and increases the loving kindness we extend to ourselves.

The exercise consists of three simple steps. First, identify ten to twenty people from different areas of your life: colleagues, employees, supervisors, friends, family, and so on. Then, send an email asking them to share two to four examples of situations where they've seen you at your best. When responses start rolling in,[11] identify the key themes (e.g., creativity, empathy, etc.). Finally, sit down and write a quick portrait of you at your best. Focus on a few paragraphs summarizing what you learned using the prompt "When I am at my best . . ." (For a "paint-by-numbers" RBS guide and sample best self–portrait, see Appendix F.)

When my clients first hear about this exercise, they often ask (with a hint of skepticism) whether this is just a "smoke-blowing" exercise—in other words, soliciting platitudes about how great we are. In response to this very fair question, I point out that the RBS is not about ego, social

---

10. Unless you want one, in which case, call me! ☺
11. As scary as it may feel to ask, you'll be surprised how eager to assist most people will be.

approval, or blindly boosting our self-esteem; it's about gaining a holistic and data-driven picture of our specific defining strengths through the eyes of others.

Another question I'm often asked is whether you can limit outreach to colleagues if work is where you need the data. While you certainly can, research suggests that a diverse mix of personal and professional connections yields the most comprehensive picture of your superpowers. This is why I always interview friends and family when assessing my coaching clients—in general, they've seen us in a broader range of situations over a longer period of time.

A final question is, "What if I don't *just* want to hear the good stuff? What if I want constructive criticism too?" To that, this self-awareness researcher says, good on you! You can indeed make that happen: a variation of the exercise is to ask for two to three examples of your greatest strengths and one to two examples showing how you could improve or grow. (Just make sure you're asking "loving critics"—that is, people who want you to succeed *and* will be willing to tell you the whole truth.) Intriguingly, one study shows that this variation may result in even *more* positive emotions than focusing on strengths alone.[12]

Sometimes, the RBS will reveal strengths you know you have. Other times, you'll discover a hidden superpower—something that others see but you've missed. Still other times, you'll find that something you regard as a weakness is actually a defining strength to those around you. For example, Mark, an army squad leader, worried that his team saw his well-intentioned candor as a lack of tact. He was stunned when one squad member shared that Mark was at his best when voicing his true perspective while offering support. Mark's compassionate candor, he learned, wasn't just accepted; it was appreciated. Gathering a broad portrait of how others see us doesn't just boost our sense of confidence, it offers new information to deepen our impact.

---

12. The researchers posited that when we "only" receive strengths-based feedback, we might try to read between the lines to see what wasn't said. As per usual, if you find yourself wondering, it's better to just ask!

ENVISION YOUR FUTURE SELF

At the end of Grace's whirlwind feedback day, I gave her some homework—something I call the **future you exercise.** This tool is particularly useful in helping replace our shadow habits with shatterproof ones, and goes something like this:

1. **Honor "past you."** Past Grace worked hard, doing anything and everything (and more) that was expected. She spent decades building relationships and exceeding expectations. It's time for "present Grace" to thank "past Grace" for her tireless efforts and to leave that part of herself—the one that constantly worries she's incompetent and letting people down—in the past.
2. **Fully see "present you."** The next step was something that didn't come easily to Grace: to start believing her own hype. It was time for her not only to hear, but truly internalize, what people were telling her—that she was already doing an exceptional job.
3. **Commit to the "you of tomorrow."** Grace chose a date to go all in on her transformation, writing a "Day 1 Manifesto" for herself to officially declare what habits and behaviors she would change. I offered a few suggestions:

Today is the day that I grant myself permission to stop:

- Proving my value.
- Being shy around important people.
- Worrying what others think about my contributions.
- Waiting for permission to recognize my own worthiness.

Today is the day I'm going to start:

- Owning the important decisions I make as CEO.
- Delegating more so I can focus on my highest and best contributions.
- Trusting my confidants to help me see what I'm missing.

Crafting Confidence

- Prioritizing my own needs in addition to being of service.
- Remembering not to be limited by anything, least of all myself.

That weekend, Grace took a solo trip to the mountains (her "happy place") to reread her report. By the time she returned to work, her colleagues' evaluations were starting to sink in. In an email to me, she shared, "It's still challenging to read the strengths section, but I am working on internalizing that feedback."

Our next step would be to help Grace deepen and sustain her budding confidence—so that she could summon it in the moments she needed it most.

Quick Competence Wins

In our next meeting, Grace admitted to feeling overwhelmed by the work ahead of us. The positive feedback had sunk in, but would she be able to approach her daily life from a place of confidence, especially in moments of unexpected stress or setbacks?

"That's a reasonable concern," I replied. "I wonder if we can find an example where you approached a huge, scary goal and blew it out of the water."

Grace thought for a moment. Then, with a nonchalance I'd soon learn was typical for her, she answered, "Well, last year, I decided I wanted to eat healthier. I ended up losing almost fifty pounds, and more importantly, I felt great."

I couldn't help but chuckle. "Well, I'd say that's a pretty good example! Let's break this down: How did you translate that huge goal into smaller steps you could take every day?"

Grace's eyes twinkled as she distilled her whole weight loss journey into a few words. "I ate more cherry tomatoes and less Doritos."

"Ah, of course! And how did you do that?"

"I stayed focused on my goal by journaling regularly. I diligently prepared, always carrying healthy snacks with me to stay ahead of my hunger. And, again in my journal, I celebrated my wins."

In the same way that bringing healthy snacks wherever she went helped Grace stay ahead of her hunger, she could forge new habits to help her starve her doubt dragon. She had already learned two important confidence-crafting strategies: **engineering, then celebrating, competence-boosting "wins."** (Conveniently for perfectionists, research shows that regular confidence crafting is a powerful strategy to distract us from dwelling on our deficiencies.)

Take, for example, Grace's desire to "stop being shy around important people." When interacting with her board, "past Grace" was timid, constantly worrying about how she was being perceived. However, after hearing the feedback on her exceptional interpersonal prowess, "present Grace" realized that the best thing she could do was to just be herself. But to avoid slipping into a timid posture around them, she adopted a new habit she called "be the host." By logging in twenty minutes early to a virtual board meeting and imagining she was greeting arriving participants as guests in her home, she was surprised at how easily she chatted up so many powerful people. Afterward, several board members reached out with glowing appreciation for her engagement.

Another one of Grace's goals was better delegation. During "feedback day," she'd admitted that she had lived in the trenches for so long that she feared stepping out of them. Impostor syndrome loomed large: "I don't want people to say, 'What the hell does she do all day?'" Eventually, Grace conceded that true success as a CEO would come from focusing on her highest and best contributions. To do that, she'd have to get comfortable letting go of the day-to-day by sharing knowledge and power with her team.

Grace began journaling her daily successes and roadblocks, just as she'd done to hold herself accountable for her healthy eating journey. A few weeks later, she reported that delegating was far easier than she expected. "I love helping other people take the lead and supporting them to be successful," she told me. "And *I* feel like I'm making a bigger impact."

The 10 Percent Buffer

About a month later, Grace shared another update. The good news was that she was calmer and more confident when things went wrong.

The less good news was that occasionally, she felt herself "sinking into despair." The reason? Not every decision she made as CEO was turning out to be universally popular. She knew that came with the territory, but didn't know how to square it with her exceptionally high standards.

It's common for perfectionists to struggle with a belief called **conditional acceptance:** the fear that even minor mistakes will result in lost respect, support, and appreciation. A related thinking style is a bias called **black-and-white thinking,** which can be summed up as, "Did I meet the standard I set? If yes, I get a gold star; if no, I'm a total failure."

One thing I can conclusively tell you about being a CEO is that, given the size and scope of a CEO's decisions, trying to keep everyone happy all the time is just about as realistic as folding the perfect fitted sheet or summarizing the plot of the movie *Inception* in one sentence (which is to say, not at all).

I asked Grace whether the belief "I've only succeeded when everyone is happy" was serving her as a CEO. Laughing, she said it obviously wasn't. So we decided to change the standard.

"What belief could you replace this with that would still feel authentic to you?"

As I'd come to expect, my client generated a gem: "If I can't make everyone happy, maybe I can make sure everyone feels heard even if they're not happy." Bingo.

This is a great illustration of a tool I call the **10 percent buffer**—permitting ourselves to be excellent 90 percent of the time—overall, a solid A-minus.[13] Shatterproof people understand that occasional misfires and mistakes occur in a greater context, and that people whose feathers may be temporarily ruffled have lots of other data, amassed over time, to consider when assessing our competence. In other words, once we've earned the benefit of the doubt, reasonable people are far more likely to extend it than we might think.

---

13. It is probably obvious that sometimes we need to set the standard at 100 percent—professions like surgery, engineering, accounting, and air traffic control come to mind.

A few weeks later, Grace put this strategy to work after one employee took to social media to voice displeasure with a decision that the larger workforce had generally welcomed. "At first, my blood pressure went up," Grace recalled. "But I took a deep breath and reminded myself I couldn't make everyone happy." By allowing that 10 percent buffer, she came to see that it was okay if some people were displeased with her some of the time. This took a lot of the anxiety away.

That year, Grace's company's revenue grew at a record pace (25 percent!), shattering financial goals across the board. Though the new CEO's job wasn't any easier than before, she felt a new sense of ease. Just as important, her team was engaged, her stakeholders were thrilled, and she was feeling more confident than ever. As Grace powerfully described it:

> Last week, I felt comfortable in my own skin. More comfortable than I'd ever felt before. I used to wake up every day with a sense of insecurity. Not anymore. Now, I'm not second-guessing myself or trying to pretend. I am who I am. The added layer of how I'm being perceived is gone. Shedding that has given me confidence and brought me a new level of happiness and contentment.

Fittingly, just as I was finishing this chapter—about nine months after we concluded our coaching—Grace reached out with exciting news. The prior week, the board and her co-CEO had convened and made a momentous decision: it was time for Grace to fully take the reins, a year earlier than they'd initially planned. Bursting with excitement, I promptly flew out to take her to lunch. Sitting across from my superstar client, I couldn't help but notice how much she'd changed. Gone was the leader who once felt the need to prove herself. In her place stood one whose quiet, calm confidence absolutely filled the room. As we raised our iced teas to celebrate her accomplishment, we weren't just marking a milestone in Grace's career; we were marking the dawn of the next chapter in her shatterproof journey.

OTHER WAYS TO CRAFT CONFIDENCE

Just as there are many ways to feel incompetent, there are many ways to reengineer confidence. The process can take different forms depending on the triggers and shadows we're facing. Let's look at how we can restore confidence in two additional situations: when other people challenge our contributions, and when we're reeling from a major setback. (As another resource, Appendix G will help you assess your current confidence satisfaction and frustration.)

Finding Our Worth When Others Come at Us

Sometimes our biggest confidence threats come from within, like Grace's did; other times, they come from our environment. How can we craft confidence when other people—especially those in positions of authority—go out of their way to make us feel incompetent?

This was the situation Jim found himself in after he was unexpectedly assigned to a new client project. Although it wasn't a perfect fit for the engineer's skill set, he was assured it would be a growth opportunity. The new project came with a new project manager—a sales guy with almost no technical knowledge who was constantly downplaying the value of Jim's expertise. And their differing views on Jim's role led to repeated conflicts, leaving him feeling like he couldn't do anything right.

Initially, Jim responded by taking on more and more work, intent on proving his value to his boss. But this shadow habit of workaholism only left him feeling more exhausted and deflated, like an invisible cog in a machine. He started questioning everything—*What am I doing here? Do I even know everything I think I do? Am I even a valid person?*—and counting the days until the project was over.

Eventually, Jim realized that he'd been seeking confidence in the wrong place: namely, his boss's approval. So instead of asking, "How many more hours can I put in?" he pivoted to "Where else can I add value?" With this in mind, he volunteered to help with a side project where his skills were sorely needed. He also began actively mentoring new hires to show them the ropes.

By finding new ways to contribute that shined a light on his true talents—rather than fixating on his boss's opinion—Jim amplified his impact, strengthened his reputation, and, most importantly, boosted his confidence. It's not a coincidence that soon afterward, the powers that be moved him into a role he loved, trading on his immense contributions from his previous project. Jim's story is a powerful reminder that **we don't need to rely on other people's respect or approval to craft our own confidence.**

Growing Forward from Major Setbacks

Sometimes, as with Jim, our confidence need is thwarted when someone challenges our contributions. But what happens when we have truly underperformed or failed despite our best efforts? When faced with our actual shortcomings—rather than our imagined ones—how can we restore our confidence *and* shore up the skills to avoid a similar fate in the future?

Juan was a talented graphic designer who was passionate about his job. But five years into his tenure, his world was turned upside down when his company was acquired. As the companies consolidated, Juan's role was shifted to the organization's sales arm, where he was asked to take on a significant amount of marketing work. Repeated attempts to explain to his new boss that he was a graphic designer, not a marketing expert, fell on deaf ears.

Ever the high achiever, Juan decided to jump in and give his new boss what she was asking for. Yet no matter how hard he tried, his work never quite measured up. He soon found himself putting in sixty- to eighty-hour weeks trying to meet his boss's standards. But for all the effort, his deliverables still fell short. Normally an energetic, optimistic person, Juan became unmotivated, creatively blocked, and increasingly bitter about the unfair derailment of his once-successful career. The day that Juan was let go, he punched his desk in a fit of rage, his fist colliding against the lacquered wooden surface with such force that his prized meditation bracelet snapped in half. (The irony was not lost on him.)

Reflecting on what had gone wrong, he eventually realized that he had responded to a new challenge with old solutions that were no longer working (not unlike Chapter 1's budworm battlers). Indeed, research shows that **the more success we've had in the past, the *less* likely we are to spot signals that we need to pivot.** Here, while Juan's optimistic attitude wasn't inherently bad, believing that everything would work out—because it always had in the past—blinded him to the reality that he needed to take more proactive action to widen his skill set. When we're gripped by inertia, we often fail to confront the question: What am I pretending not to know?[14]

Rising above his self-pity, Juan asked himself, "What can I do to advance my career and make sure this never happens again?" He decided on a multipronged approach, investing a third of his severance check to enroll in a series of intensive marketing courses, another third to move to London to take a job at an advertising agency, and the rest to begin cognitive behavioral therapy to manage the emotional fallout of the experience.

While it would have been easy for Juan to jump back to graphic design and blame his boss for putting him in the wrong role, he believed that there was more ahead of him by continuing down the path he was on. Today, he still works in marketing and hasn't touched graphic design in five years. In his second career, he no longer sees constructive feedback as a sign of inadequacy but as a well-meaning invitation for growth. For example, when his new boss told him he needed to write better briefs, he was miffed at first—but quickly saw the opportunity to deepen his shatterproof goal of mastery. He immediately enrolled in two workshops, which ultimately helped him and his team craft better, more creative, award-winning work.

More broadly, Juan's transformation has helped him, as he says, "remove some of the textbook expectations I had about cause and effect. I've come to terms with the reality that improving X doesn't guarantee Y."

---

14. I first heard this from my friend Chuck Blakeman, and it's become one of my most frequently asked self-reflection questions.

This has given him a deeper sense of peace, even when things don't go his way.

As Juan, Jim, and Grace each learned, when we take an active role in managing our self-doubt—by discovering our strengths, applying our skills in new areas, and investing in our personal and professional growth—we'll feel more confident in our present performance and be ready for future challenges.

CHAPTER 9: KEY TAKEAWAYS AND TOOLS

"No matter what obstacles we're facing, confidence-crafting opportunities are all around us if we just know where to look."

- **Confidence versus self-doubt**: Feeling like we're doing well and getting better is surprisingly unrelated to our actual abilities—self-doubt festers in the face of triggers like expectations, chaos, and criticism (reference page 103).
  - **Sample confidence shadow goals**: Achievement, defensiveness, inertia (see the table at the end of this chapter).
  - **Sample confidence shadow habits**: Paranoia, excessive self-focus, unnecessary competitiveness (see the table at the end of this chapter).
- **Taming our doubt dragon**
  - **Terms**
    - **Impostor syndrome**: Feeling incompetent despite evidence we're doing well.
    - **Metaperception**: Our perception of how others see us.
    - **Black-and-white thinking**: Mistaking a lack of perfection for failure.
    - **Conditional acceptance**: Believing even minor mistakes will cause others to reject us.
  - **Tools**
    - **Reflected best self exercise**: A process to get feedback to learn how others *really* see us (see Appendix F).

- **"Future you" exercise**: Marking your pivot by honoring "past you," appreciating "present you," and committing to "the you of tomorrow."
- **The 10 percent buffer**: Whenever possible, giving ourselves permission to be excellent "only" 90 percent of the time.

- **Finding confidence when others challenge it**
  - Remember that we don't need other people's approval to craft confidence.
  - Find new ways to add value that don't depend on naysayers' opinions.
- **Growing forward from major setbacks**
  - Don't let past successes close you off to early signals you need to pivot.
  - Most feedback is a well-intentioned invitation for growth.
  - When you feel stuck, ask, "What am I pretending not to know?"

Thwarted Confidence: Common Shadow Goals and Habits

Category	Shadow Goals	Shadow Habits
Protect: Fighting back to avoid guilt, shame, or blame.	• **Defensiveness**: A drive to protect ourselves against criticism.	• Overreacting to feedback, taking things personally. • Quickly dismissing feedback that challenges our self-concept. • Doubling down on clearly ineffective strategies. • Downplaying mistakes. • Deciding that everyone thinks we're awful despite evidence to the contrary. • Blaming others for our flaws and insecurities. • Seeking superiority over others, often by belittling them.
Prove: Seeking external evidence about ourselves or our place in the world.	• **Achievement**: A drive to achieve greatness at all costs. • **Rewards**: A drive to amass material rewards like wealth or image. • **Perfection**: A drive to meet excessively high standards.	• Excessive focus on ourselves or our performance. • Compulsive focus on winning. • Excessive competition or social comparison. • Obsessive passion for work (i.e., workaholism). • Frequent self-criticism and negative self-talk. • Believing that mistakes mean we're a failure. • Paranoia. • Seeking constant praise. • Obsession with impressing others. • Believing our worth must be earned.

Shatterproof Transformations and Tools

Category	Shadow Goals	Shadow Habits
Prevent: Denying experiences, or devaluing our needs, to feel better right now.	• **Escape**: A drive to avoid situations that hurt our confidence. • **Inertia**: A drive to maintain the status quo. • **Ignorance**: A drive to ignore difficult (self-) truths.	• Self-handicapping behaviors. • Procrastinating on important tasks. • Seeking distraction over contribution, or activity over results. • Choosing easy goals or activities over more ambitious ones. • Being trapped in the comfort of our preferred routines. • Opting out of situations that push us outside our comfort zone. • Assuming that past achievements guarantee future ones. • Allowing ourselves to be paralyzed by fear.

CHAPTER 10

Crafting Choice

"The most common form of despair is not being who you are."
—Søren Kierkegaard

Nestled amid the majestic snowcapped peaks of the Dinaric Alps, the sun-kissed coastlines of the Adriatic Sea, and the turquoise waters of Uvac Canyon lies a region rich in natural splendor and cultural diversity. Yet beneath the Balkans'[1] surface beauty is a turbulent history that's earned it the title of "Europe's powder keg."

At the heart of this complex narrative stands one man. Unlike *Harry Potter*'s archetypal villain, Voldemort—to whom he's been compared—Slobodan Milošević's personal history doesn't exactly scream "future genocidal dictator." After studying law and settling into domestic life with his high school sweetheart, he began a career as a business administrator. Later in life, as a budding politician, the balding, dough-faced, big-eared ex-banker was generally regarded as wooden and unremarkable. But Milošević's star quickly rose thanks to the tensions he stoked with neighboring ethnic groups, and in 1990, he rode a wave of xenophobic nationalism right into Serbia's presidential palace.

You may recall what happened next. The "Butcher of the Balkans" entered into bloody conflict with neighboring Slovenia, Croatia, Bosnia and Herzegovina, and, later, Kosovo, where his forces perpetrated

---

[1]. Located in Europe's most southeastern peninsula, the region includes the former Yugoslavian republics of Croatia, Montenegro, Serbia, Slovenia, Bosnia and Herzegovina, and Macedonia, as well as Albania, Bulgaria, and Romania.

especially horrific atrocities against Bosnia's Muslims and Kosovo's Albanians—torching villages, building concentration camps, and systematically raping tens of thousands of women. In total, the dictator is responsible for 125,000 deaths, the displacement of three million people, and the collapse of the Yugoslav federation.

Within Serbia's borders, too, life was bleak. The nation's seven million citizens were subjected to a regime of repression straight out of the authoritarian playbook. Milošević murdered political adversaries, jailed activists, and rigged elections, employing every means to deprive Serbia's citizens of their agency and autonomy. Meanwhile, government corruption and economic mismanagement plunged Serbia into unprecedented poverty.[2]

It was under this backdrop that Srdja Popović, then a first-year biology student at the University of Belgrade, would experience a political awakening. The second son of two TV reporters, Srdja grew up dreaming of making animal documentaries like his idol David Attenborough. At university, he split his time between coursework, playing bass in a goth band, and hanging out with his friends.

On the surface, Srdja and his buddies were typical college students—they liked drinking beer, staying up late, and trying to score dates. But beneath the surface, they shared something deeper: an abhorrence for Slobodan Milošević and everything he stood for. Like most Serbians, though, Srdja and his friends felt helpless in the face of his total oppression.[3] After all, challenging a genocidal tyrant with his own police force seemed like a pretty stupid move.

---

2. Unemployment levels reached 50 percent, and in just one year, the price of two pounds of potatoes surged from 4,000 dinars to an incredible 17 billion (Steve York, dir., *Bringing Down a Dictator* [Washington, DC: International Center on Nonviolent Conflict, 2002], https://www.nonviolent-conflict.org/bringing-dictator-english); Christopher Hitchens, "No Sympathy for Slobo," *Slate*, March 13, 2006).

3. Others responded by fervently supporting the very regime that was stealing their freedom. Research shows that in the face of choice thwarting, some people respond by paradoxically backing controlling institutions (Aaron C. Kay et al., "Compensatory Control: Achieving Order through the Mind, Our Institutions, and the Heavens," *Current Directions in Psychological Science* 18, no. 5 [2009]: 264–68).

That is, until May 1998, when parliament passed the University Act. With this de facto government takeover of Serbia's six universities, regime-appointed deans swapped employment contracts with loyalty oaths, purged esteemed faculty, and replaced curriculum with propaganda. This new development threatened students' freedom and future in a uniquely frightening way—one that Srdja and his friends could no longer ignore.

With no savior in sight, they faced a choice: either sink deeper into surrender or take a stand. So, one October evening shortly after the start of the semester, Srdja and a half dozen friends gathered in a cramped, smoky Belgrade apartment and founded Otpor! (Serbian for "resistance"), a movement to fight Milošević's repressive regime.[4]

Their first order of business was creating a symbol to rally behind: something bold, edgy, and, above all, *cool*. Designed by Srdja's friend Duda—to impress one of the women in the group, no less—Otpor!'s logo depicted a bold clenched fist reaching toward the sky. A few weeks later, under the cover of rain and darkness, they spray-painted three hundred fists near Belgrade's Republic Square. The city awoke to a newly defiant landscape, with each image boldly asserting that a movement was brewing. Soon, flyers with pithy slogans—"Bite the system!" "Live resistance!" "Because I love Serbia!"—were plastered seemingly everywhere.

Their campaign achieved its intended effect of attracting other young people to the movement. After weeding out the "posers" and "potential police informants," Srdja and his cohort recruited a core group of about a dozen students in their teens and twenties. They were united by their opposition to Milošević and the belief that they could build a better future for their beloved country.

THE DICTATOR NEXT DOOR

While most of us will (hopefully) never be stripped of our basic freedoms under an evil dictator, we've all had our agency thwarted at one point.

---

4. Initially focused on challenging repression at Serbia's universities, the students quickly realized that removing Milošević entirely was their only path forward.

And whether powerlessness stems from living under a dictator's thumb or more mundane frustrations, the triggers are often similar: **suppression** of our authentic selves (internal or external pressure to act against our true nature), **coercion** (external factors forcing us to obey, like micromanaging bosses or controlling parents), **loss** (losing something we value, like a relationship or promotion), **disregard** (being minimized or invalidated), **unfairness** (being the victim of inequity, like a colleague stealing credit), or **voicelessness** (a lack of transparency or input).[5] (For a refresher on choice triggers, see page 103.)

Put simply, we humans are wired to want to make our own choices, authentically express ourselves, and live in line with our values, interests, and needs—an instinct that beautifully explains every revolution and quest for freedom in human history (and why so many people love the movie *Braveheart*). This is by design: fighting coercive threats ensured our ancestors' survival.

At their core, all choice triggers exist on a continuum: on one side lie decisions springing from our true, authentic selves and on the other are those that come from internal or external pressure to act at odds with who we are and what we want.[6]

Pressure  Authenticity

---

5. (Not so) fun fact: On average, public sector employees tend to experience more choice frustration than those in the private sector due to limited resources, public scrutiny, regulatory constraints, political interference, and more bureaucratic structures (Aiste Dirzyte, Aleksandras Patapas, and Dovile Zidoniene, "Employees' Personality Traits and Needs' Frustration Predicts Stress Overload during the COVID-19 Pandemic," *Scandinavian Journal of Psychology* 63, no. 5 [2022]: 513–21).

6. If you're wondering whether the need for choice is constant across cultures—including those that emphasize collective harmony over individual empowerment—it indeed is. This underscores an important nuance about what choice is. It is less about independence and more about being able to make one's own decisions, which can coexist with relationships and social structures (Martin F. Lynch, Jennifer G. La Guardia, and Richard M. Ryan, "On Being Yourself in Different Cultures: Ideal and Actual Self-Concept, Autonomy Support, and Well-Being in China, Russia, and the United States," *Journal of Positive Psychology* 4, no. 4 [2009]: 290–304; Shi Yu, Chantal Levesque-Bristol, and Yukiko Maeda, "General Need for Autonomy and Subjective Well-Being: A Meta-Analysis of Studies in the US and East Asia," *Journal of Happiness Studies* 19 [2018]: 1863–82).

Imagine, for example, being invited to an office happy hour at the end of an especially exhausting week. All day, you've been looking forward to your pajamas and TV, but now you feel pressured to join the team bonding. Do you make the authentic choice and politely decline, or do you cave and postpone your Bravo marathon for another night?

One way to learn where a decision falls on the pressure-authenticity spectrum is simply to ask, "How do I *really* feel about doing this?" Answers like empowered, aligned, purposeful, energized, and fulfilled suggest more authenticity, while answers like frustrated, helpless, controlled, annoyed, scared, or obligated suggest more pressure.

Authenticity means being guided by a deep sense of self, which means doing things because we *want* to and not because we *ought* to or *have* to. When you're living authentically, you'll feel like the architect of your life. What's more, you'll experience the inner awareness and alignment that come from synching your actions with your values and needs. Acting authentically means never having to suppress or override your natural instincts—this not only conserves mental energy[7] but also powers well-being. When we act based on what truly matters to us, we're more engaged, creative, and fulfilled and can be of greater service to others.

On the other side of the spectrum, pressure causes stress, unhappiness, anxiety, and even depression. Studies show that the longer we yield to pressure, the harder it becomes to harness the motivation to restore our autonomy. Learned helplessness sets in, making us more likely to adopt shadow goals that give us a false sense of control or power, either over ourselves (like imposing **restrictions** on things like eating or exercise) or others (like **dominating** our team). Or we might recklessly defy rules to protect ourselves from the shame of helplessness (**rebellion**). Or we can default to what's expected of us even when it isn't in our authentic best interests (**harmonizing**), or give up on our authentic needs and choices altogether (**giving up**). (For a list of shadow goals commonly associated with choice frustration, flip back to page 117.)

---

7. Pretending to be someone we are not is freaking exhausting!

For individuals, choice shadows have costs, including boredom, apathy, and exhaustion. Collectively, these costs stifle participation, silence voices, and stop growth. (In organizations, this translates to declines in performance, and spikes in mistakes, absenteeism, and even unethical conduct.)

In the dark days of Milošević's reign, most Serbians—whose existence came about as close to the pressure side of the spectrum as you can imagine—had two options: they could fight or they could flee. Tens of thousands escaped to nearby countries to avoid being drafted into the dictator's arbitrary war. But for millions of Serbs, the only escape was mental. In Chapter 5, we explored how people often cope with extreme stress by disconnecting from their emotions—specifically, our nervous system will shield us when reality is painful enough. Given the Serbian people's persistent trauma of life under a coercive dictatorship, it's perfectly understandable that many responded not with fight-or-flight, but with freeze-or-faint reactions like denial, acceptance, and apathy.

Thankfully, Srdja and his Otpor! comrades made a bold and different choice: to fight back. And if they were going to challenge a genocidal dictator in a battle for Serbia's future, they knew they'd have to move the country's citizens from a place of fear and apathy to one of courage and unity.

The Art of Bully Jujitsu

One spring night in 1992, the Serbian rock band Rimtutituki hit the streets to promote their newest single. Unable to get official permission for a concert, the popular antiwar rockers opted for the next best thing. They would perform their tunes while cruising through Belgrade in a truck trailer. Circling Republic Square belting out such lyrics as "there's no brain under that helmet," the band resembled generals more than rockers. Their act was cheeky and subversive and not exactly subtle. Young Rimtutituki superfan Srdja Popović was profoundly affected by the experience. That night, he learned that the best resistance isn't dull or obtuse; it's accessible and, above all, entertaining.

Under choice threat, people feel caught between two extremes—either do nothing or choose costly confrontation. But shatterproof people like Srdja find a third option—an approach he calls "delegitimizing the oppressor," and I call **bully jujitsu.** If you're not familiar with the Japanese martial art of jujitsu, it's about using your opponent's energy against them (no matter their size or strength) without causing undue harm.[8] For example, you might neutralize your opponent's aggression by redirecting a punch that throws them off balance.

Srdja and his peers stumbled upon this strategy almost by accident. During an early brainstorming session, Otpor! members found themselves drowning in despair. How could they hope to rally their fellow Serbians to their cause with Milošević's fear machine in full swing? Then, someone offered a brilliant insight: "The only thing that can beat fear is laughter." That's when they dreamed up the smiling barrel.

At once, they procured a heavy metal barrel from a nearby construction site and tasked Duda, Otpor!'s logo artist, with sketching a sinister, grinning caricature of Milošević on its face. With the addition of the words, "Smash his face for just a dinar,"[9] their masterpiece was complete.

The students took the barrel and a baseball bat to Belgrade's main pedestrian boulevard, placed them in the middle of the street, and eagerly retreated to a nearby coffee shop to see what would happen.

Passersby were initially puzzled, but curiosity soon gave way to amusement. Then, a breakthrough: a young man dropped a coin in the barrel, grabbed the bat, and took a hard swing at Milošević's face. *CLANG!* The sound reverberated for blocks, but he was not deterred. Before long, others joined the barrel-bashing frenzy, recognizing the acceptably low risk of participating. As laughter filled the air, Srdja and his comrades sipped espressos and puffed cigarettes in delight.

---

8. I can speak to this with some lighthearted authority. I went to a school that had a sports requirement—but as a theater nerd, I did not think options like field hockey and lacrosse felt like "me." Instead, I took circus class and earned a blue belt in martial arts—both of which actually ended up being pretty fun.

9. The equivalent of US$0.02.

But the best was still to come.

Minutes later, a squad car arrived. Two policemen soberly began questioning witnesses, itching to identify the organizers, who were nowhere to be found. Realizing that they couldn't just haul off these joyful young people and families without sparking a public outcry, the authorities opted for the only remaining option: *they arrested the barrel.* The scene, straight out of *Monty Python*, was immortalized in widely circulated photos depicting Milošević's feared police force as a gaggle of bumbling fools.[10]

Beneath these irreverent antics was a serious strategy to dismantle fear, the cornerstone of autocratic rule. With that day's strategic experiment, Otpor!'s bully jujitsu masters boosted the country's morale—*and* incited a clumsy, embarrassing response from the regime. Indeed, **in the darkest of times, humor doesn't just neutralize fear, it gives us the courage to act.** What's more, even the smallest action to assert our rights and freedoms can deliver big results.

Choosing Choice Support

As I've gotten to know Srdja, I've learned that he is almost comically modest. After cofounding Otpor!, for example, he shouldered the massive task of training its seventy thousand members across 130 branches. In a self-deprecating nod to military-style conformity, he calls his role "ideological commissar"—but that couldn't have been further from the truth.

A cornerstone of Srdja's leadership was the shatterproof practice of **choice support.**[11] While pressure-based leaders restrict followers' choices

---

10. One more example of Otpor!'s bully jujitsu is too funny to leave out. Activists in the provincial town of Kragujevac took white flowers—representing the dictator's despised wife, who wore a plastic one in her hair every day—and stuck them on the heads of turkeys (a bird whose Serbian name is a crude insult). No one that was there would ever forget the sight of the policemen tripping all over themselves trying, and failing, to capture every last turkey.

11. This concept is called autonomy support by self-determination theory researchers, but I'm keeping the term consistent for simplicity.

in an effort to get them to do what they want, choice-supportive ones remind them, "Ultimately, it's up to you." Whether or not you're a formal leader, virtually anyone—bosses, educators, coaches, therapists, parents, siblings, friends, partners—can help others find new agency. Choice support leads to greater need satisfaction, motivation, growth, and performance. And by taking a page from Srdja's playbook, anyone can create choice for others with choice-supportive behaviors like validating their personal experiences, offering information to support authentic decision-making, and emphasizing empowered action.

Starting with validating people's experiences, Otpor!'s leaders went out of their way to normalize members' fear. Fear is a natural response to oppression, Srdja explained, but it's never the oppressor's goal ("a dictator isn't interested in running a haunted house"). Instead, fear exists to make you obey. And while there may be a cost for choosing not to obey, Srdja emphasized that whether we do will *always* be up to us.[12]

As you probably remember from Chapter 1, humans have a profound fear of uncertainty, one on which Milošević's police capitalized with the threat of arrest. Since no one in Otpor! knew what being arrested would actually be like, they defaulted to their worst imagination. Luckily, Srdja understood two things: the unknown is usually far scarier than reality, and the best way to shrink fear is to replace uncertainty with knowledge.[13] So as soon as their members started getting arrested, Otpor! leadership began documenting every detail and educating everyone on what to expect. While this knowledge couldn't completely eliminate fear, it went a long way to help people take smarter risks. Soon, many were speaking about the prospect of their own arrest with confidence, even bravado.

---

12. Consider how different this is from the stereotype of a "strong" leader telling you not to be afraid in the first place.

13. In one study, when researchers thwarted mahjong players' choice need, they found that players could regain their agency, but only when they felt subjective competence in playing the game—a testament to the importance of understanding our situation and available options (Rémi Radel, Luc Pelletier, and Philippe Sarrazin, "Restoration Processes after Need Thwarting: When Autonomy Depends on Competence," *Motivation and Emotion* 37 [2013]: 234–44).

Shatterproof Transformations and Tools

Of course, Otpor! leaders could never fully anticipate every outcome. Everyone knew that some people would lose jobs, go to prison, be tortured, or even be killed. But they were reminded that the choice to act (or not) was always theirs and theirs alone. Despite, or perhaps because of, these risks, getting arrested soon became a badge of honor, a development that Otpor! smartly harnessed by distributing shirts boasting members' arrest records—this unexpected fashion statement soon became the hottest commodity in town. Even as Milošević demanded more arrests from the perch of his grand palace, police on the ground were starting to realize that the more they cracked down, the bigger the protests were getting.

Then on January 13, 2000, thirty thousand Otpor! members and supporters gathered in Republic Square, declaring, "This is THE year that life must finally win in Serbia!" And they would do it by defeating Milošević at the ballot box. In past elections, Serbia's eighteen political parties did much of Milošević's work for him by tearing each other apart. But through the painstaking work of uniting these opposing groups, Otpor! again changed the game. By that point, all their new coalition needed to do was stay disciplined.

Deciding to call their bluff, Milošević moved the elections up ten months. One can only imagine how shocked the dictator must have been when all eighteen parties rallied behind a single candidate, Vojislav Koštunica, a constitutional scholar untainted by problematic political ties.

Fast-forward to September 24, 2000: Election Day. Thirty thousand volunteers got to work monitoring ten thousand-plus polling places to prevent fraud. Nearly 90 percent of voters under age twenty-nine took part, and nearly all voted for Koštunica. By the day's end, monitors announced Milošević's resounding defeat. Otpor! had finally toppled the dictator—but would he go quietly, or at all?

To force Milošević to step down, the opposition coalition initiated a series of civil actions. The coal miners were first to strike; then sector after sector ground to a halt while ordinary people showed up to protect strikers. Next came school boycotts, then blockades, and then public transport workers parking buses and streetcars in the middle of major intersections and taxi drivers forming slow-moving traffic barriers.

By October 5, the entire country was at a standstill. After years of work and countless small wins, it was finally time for the big rally.[14] Hundreds of thousands of peaceful protestors poured into Belgrade chanting, "Serbia has risen!" With a few exceptions, the police acknowledged the palace's orders to break up the protests but did not obey them. Even Milošević's top commanders refused to order their troops to fire. Why? Their own kids were in the crowd.

In *Harry Potter and the Half-Blood Prince*, Albus Dumbledore notes, "Have you any idea how much tyrants fear the people they oppress? All of them realize that one day, amongst their many victims, there is sure to be one who rises against them." The Serbian people had finally risen up, leaving their feckless dictator no choice but to concede. Kostunica was sworn in as president, and following a thirty-six-hour standoff, Milošević was arrested for genocide and crimes against humanity. He was tried by the International Criminal Court for war crimes, but in 2006, he was found dead in his cell, never to see the trial's completion.

Since then, the path forward for Serbia has been challenging. But the lesson remains: even when we feel beaten down, powerless, or like we have no options, **we can always choose to fight for positive change and rally others to help us get there.** If Srdja and his college buddies could create choice for seven million citizens under a murderous dictator, the rest of us can certainly craft a little more choice in our own lives.

This was a lesson I was about to learn firsthand.

THE 2-2-2 TOOL

As I strode through the automatic doors of the world-renowned EDS clinic, along with my color-coded binders and a newfound sense of agency,

---

14. Srdja explains that contrary to what most people think, the big rally is the victory lap for successful movements—not the start of them.

I was optimistic. After hitting countless dead ends in my search for a diagnosis, I was certain that these specialists (whom I'd flown across the country to see) would have the answers. And with these answers, I was certain, would come more choices and chances to feel better.

I couldn't have been more wrong.

Fifty-five minutes later, I stumbled out of the medical center into an impeccably manicured park full of smiling patients and visitors and promptly started hyperventilating.

How could this happen? To my utter astonishment, it was one of the worst medical experiences I've ever had, and that's saying something. The head of the clinic did a cursory exam, discounted several objective signs of EDS, and diagnosed me with fibromyalgia—an autoimmune disease that doctors had already ruled out. Then he offered to send me to a seminar promising to teach me how to "ignore pain." (You just can't make this stuff up.)

I shakily walked three blocks to my hotel, opened the door to my room, and dove under the crisp bedsheets. I was officially out of options. What on earth was I going to do?

Then I remembered a tool I was working on for this book. When strategic experiments don't go as planned, we often feel pressure to create a Plan B immediately and continue the fight. But shatterproof people tend to opt for a very different strategy: a deliberate timeout. **When we're feeling broken by the pressures of life, sometimes we need to let ourselves wallow a little before we can pick up the pieces—especially when we're running low on resilience.**

The **2-2-2 tool** permits us to pause for exactly forty-eight hours—no more, no less—while prioritizing what we need in the next *two minutes*, *two hours*, and *two days* to regain our fighting spirit. My answers were a good cry, an early bedtime, and giving up—for exactly two days. Faced with yet another disappointing setback, I used this tool to make a few small but meaningful choices and regain a semblance of control over my circumstances. I still wasn't sure of my next steps, but for the next forty-eight hours, I didn't need to be.

OTHER WAYS TO CRAFT CHOICE

As with confidence, our perception of our own agency doesn't always align with the objective reality. What's more, in the face of disappointments and setbacks, motivated, ambitious, high achievers—those with the highest expectations of controlling our destiny—tend to struggle the most. **The greater our expectation for choice, the more jarring it is when that choice is thwarted.**

It's natural to think that we could regain control if we simply had more formal power or influence. But my work has taught me that even the most powerful people are bound by forces outside their control. In organizations, for example, employees feel constrained by managers, who feel constrained by executives, who feel constrained by the C-suite, who feel constrained by the CEO, who feels constrained by the board, who feel constrained by shareholders, and on and on it goes. (You can play this fun game with any system, from families to schools to communities.)

So if power isn't the answer, how else can we successfully navigate situations that limit our agency and create more choice in our lives? Such transformations can take many shapes depending on the triggers and shadows involved—but no matter how much our environment restricts our autonomy, the following strategies can help us fulfill this need in new, creative ways. (To assess your current choice frustration and satisfaction, see Appendix H.)

Coming Back from Rock Bottom

Looking back, Gerone sees his father-in-law's diagnosis of stage 4 lung cancer as the moment his life started falling apart. While his exhausted wife shuttled between Arizona and California to care for her father, Gerone was left to handle all the household responsibilities while supporting their family through his demanding job as a VP of operations. When his father-in-law passed away just before Christmas, Gerone was heartbroken but hoped that the worst was behind his family.

He was wrong.

Weeks later, he received a concerning call from his mother in Mississippi: his father had vanished during a solitary walk. The chilling discovery of their beloved dad's lifeless body hanging from a tree left Gerone reeling as he struggled to comprehend why his hero had died by suicide.

Then, Gerone's son was involved in a motorcycle accident on his way to work. Fortunately, he recovered, but not before spending weeks in the hospital and saddling the family with astronomical medical bills.

Beyond Gerone's grief for his father and father-in-law, concern over his son's injuries, and stress over his mounting debt, he couldn't help but feel broken by the unfairness of it all. *Why me?* he wondered. *What have I done to deserve such misfortune?* Clearly, the universe was hell-bent on thwarting his every attempt to get his life under control. People tried to comfort him, saying there was nothing he could have done to avert these disasters, but this only made him feel more helpless.

As if all this loss and misfortune didn't cut deep enough, Gerone's professional life was also unraveling. Looking on in horror as his key metrics—quality, on-time delivery, employee retention—plummeted, the VP's stress levels reached record highs while the organization's trust in his team reached record lows.

The thought of digging out felt like "trying to solve a jigsaw puzzle in a category 5 hurricane." Gerone contemplated his father's tragic fate, finding a chilling parallel to his own experience. Maybe his father had ended his life because he saw no way to restore control. For the first time, Gerone understood how his father must have felt. This wake-up call motivated Gerone to act.

First, he asked himself, *Of all the things causing me pain, which can I control?* Then, he took a personal inventory. He couldn't change the market dynamics hurting his metrics; he couldn't print money to pay his bills; he couldn't know the real reason his father had ended his life. What could he control? Only one answer stood out: "My body," he realized, "is *mine*." He chose to prioritize his health—not because he *wanted* to, but because he *had to* and he *could*.

The process began with an overhaul of his eating habits. Next, he started regular walks, which turned into half-jogs and, before long, fat-burning, muscle-building runs. Once he'd mastered these habits, Gerone started strength training. Within nine months, he had shed a whopping one hundred pounds.

And what he lost in weight, he gained in motivation. Finding just one area to craft choice around allowed a sliver of bright light to shine through Gerone's heavy cloud of fear. Because three-to-thrive needs are fulfilled holistically, research shows that **need crafting in one life domain improves need satisfaction in others.** Gerone was no exception: soon, he was seeing opportunities for change everywhere. At work, he helped his team narrow their view to things they could control, shifting paradigms and finding new ways to make the market weirdness work.

That year, Gerone's team delivered the best metrics in the company's fifty-year history. Meanwhile, while training for his first half marathon, he'd never felt more at ease in his body or his mind. After hitting rock bottom, Gerone didn't just bounce back to where he had been before; he grew forward. Two years later, he's sustained and improved his gains at work and is now competing in ultramarathons. This is living proof that when we dare to craft choice—even in just one area of our lives—the ripple effects can be remarkable. (For a list of common shadow habits for choice, see page 179.)

Choosing Authenticity over Mustivation

Scott felt lucky to have found his calling early in life. At least, that's what he thought, until the day he woke up to realize that after years of hard work—including the doctorate he'd been so proud to earn—his passion had become the bane of his existence. Isolated, burned out, and broken, Scott agonized, *Have I dedicated my entire career to something that will never make me happy?*

Searching for reasons for his job dissatisfaction, he found a company culture that paid lip service to well-being while demanding employees "do more with less" (such disregard for our time, workload, and well-being is a common workplace trigger). But sensing there was more to the situation, he dug deeper. Over the years, he'd voluntarily taken on countless commitments beyond his job description—conference presentations, consulting, blogging, volunteering, and so on, convincing himself that he was doing them out of genuine interest. But after probing how he *really* felt, Scott realized he was doing them out of mustivation more than motivation.

When our choice need is frustrated, we tend to blame external forces stealing our agency and power—as Scott did initially. But as he discovered, choice triggers don't always come from bosses or colleagues or the company; they can also come from within. Here, Scott had internalized ideas about what "good" professionals did, leading to the shadow habit of putting others' expectations over his own needs.

As we go through life, we build up assumptions and expectations about who we are. This sense of self—our **identity**—is generally shaped by the groups we belong to ("I'm an Apple employee"), our roles ("I'm a parent"), and our individual characteristics ("I'm a good communicator"). All three anchors help us maintain psychological equilibrium during hard times, but the groups and roles we identify with offer particular safety because of the social support and shared purpose that come with them.

Having a role or group become central to our identity is riskier than we might realize. Although strong connections with our company and colleagues drive positive personal and organizational outcomes, we can go too far, and **overidentification** is detrimental. Being too intertwined with a certain role or group leads us to lose perspective and eventually neglect our own needs. Unfortunately, this shadow habit is all too common. Some organizations even encourage it by promoting a "company first" culture, rewarding employees for personal sacrifices and putting the company over their well-being.

There's a reason researchers label overidentification "a pathology": it fosters workaholism, poor well-being, and loneliness,[15] among other

---

15. This state was described by William James as *Zerrissenheit*, German for "torn-to-pieceshood" (Elizabeth Glendower Evans, "William James and His Wife," *Atlantic*, September 1929).

outcomes.[16] Even worse, when our job becomes our identity, setbacks or losses at work can trigger anxiety, depression, anger, humiliation, and even PTSD. As psychologist Janna Koretz writes in a pointed *Harvard Business Review* article, "Hating your job is one thing; but **what happens if you identify so closely with your work that hating your job means hating yourself?**" (See Appendix I for an exercise to see whether you're overidentified.)

You might assume that companies would still benefit from having fleets of foot soldiers who believe the company's success is their own. But overidentification is harmful here too, in the form of decreased creativity, smart risk-taking, and performance. It also fosters conflict, resistance to change, and makes employees less likely to speak up against unethical behavior. The consequences of overidentification are especially problematic when leaders have to make tough decisions. What happens to the "we're a family" narrative after a round of layoffs?

Scott realized that his career had become the anchor of his identity, leading him to pressure himself to suppress his authentic desires.[17] It was stressful and draining. No wonder he was hating his life's calling! Then and there, he vowed to build a more balanced identity, no longer allowing any one element (career, company, family, marriage, parenting, community, leisure, etc.) to have a disproportionate influence over who he was.

His first strategic experiment would be to start saying no. Often, we agree to things we don't want to do because we feel we don't have a choice.[18] But we can overestimate the consequences of turning down

---

16. I would like to thank the wonderful beta reader who noted that "the same thing can sometimes happen to parents who pour [energy] into their children at the expense of the rest of their lives once the child leaves the house."

17. This notion of sacrifice—excessively doing things we don't want to do in service of our career or our future—is a particular problem in the United States (Anne C. Holding et al., "Sacrifice—But at What Price? A Longitudinal Study of Young Adults' Sacrifice of Basic Psychological Needs in Pursuit of Career Goals," *Motivation and Emotion* 44 [2020]: 99–115).

18. This precisely is why being too busy makes us miserable. (See Cassie Mogilner, "It's Time for Happiness," *Current Opinion in Psychology* 26 [2019]: 80–84.)

requests, meaning we may have more control in these situations than we think. As author Arthur Brooks advises, "You just have to get the knack for saying no instead of yes." There are lots of ways to say no, including putting off the decision, asking for a different deadline, or offering up someone else who can help. We rarely have agency over all of our choices, but we must exercise it wherever we can.

Scott decided to drop the extracurriculars that didn't bring him joy or add value to his life or his work, noticing how freeing it was to let them go. Where he used to pride himself on soldiering through his endless commitments, he now found joy in escaping so much unnecessary, self-imposed stress. Moving forward, whenever he received an extracurricular request, he would pause and remind himself to say no to anything that didn't align with his values and interests.

At the same time, Scott prioritized the things he cared about outside his job: parenting, marriage, leisure activities, and so on. He started spending more time with his family, reading more fiction, and even pursuing a small side gig. In so doing, Scott discovered that his true identity was much fuller and richer than he had ever realized. (See page 179 for more ideas on building a balanced identity and a list of need-supportive leisure activities.)

From this chapter, I hope that one thing has become abundantly clear: **no matter how pressuring or controlling our environment is, we can always choose to live more joyfully and authentically.** Or as American poet Ella Wheeler Wilcox reminds us, "There is no chance, no destiny, no fate, that can circumvent or hinder or control the firm resolve of a determined soul."

Three days after I gave myself a forty-eight-hour break in my search for a diagnosis, as if on cue, a message came from the office of Dr. Schofield (the EDS specialist I was told I'd have to wait fifteen months to see). An appointment had become available in six weeks. Trying to push aside the sense of weary foreboding, I once again began the process of updating my medical summary and preparing my four large binders of medical records and research.

Crafting Choice

The morning of the appointment, I did my best to put myself together in a form that resembled a human woman, stacked my binders in a rollaboard suitcase, and drove the twenty minutes to Dr. Schofield's office.

After the appointment, Dr. Schofield's assistant, Chandra, asked how it went. "I'm . . . I'm rarely at a loss for words," I stammered. After an uncomfortably long pause, I finally choked out, "I think you guys just saved my life."

Chandra solemnly shook her head. "No. *You* did. Our patients spend their entire lives in pain while being dismissed by their doctors and loved ones. Please don't underestimate what it took for you to get you here." I hugged her, packed up my binders, and wheeled them back to my car in the freezing winter air.

I drove home that day with two definitive diagnoses and a third confirmed by labs a week later: hypermobile EDS, postural orthostatic tachycardia syndrome, and mast cell activation syndrome. And Dr. Schofield assured me these weren't borderline diagnoses. They were clear, objective, indisputable findings.

Of course, this wasn't the finish line, nor did I have any illusions about the journey ahead of me. But here's what I could be sure of: I had new agency and a new path to follow—one that would eventually help me feel better than I had in years. (There's no treatment or cure for EDS, but you *can* treat its co-occurring conditions.) What's more, I was now building a support system of caring medical professionals and a new community of fellow EDS warriors that would become an immense source of wisdom and strength.

As I drove home, I couldn't believe the variety of emotions coursing through me. I felt vindicated, even a bit smug. I mean, with no medical training, I *had* solved a mystery that had stumped specialists for decades. But when those ego-driven emotions fell away, I was left with a feeling unlike anything my resilience practice had ever given me: a profound sense of clarity, confidence, and peace.

In that moment, I finally understood what it felt like to be shatterproof.

CHAPTER 10: KEY TAKEAWAYS AND TOOLS

"No matter how pressuring or controlling our environment is, we can always choose to live more joyfully and authentically."

- **Authenticity versus pressure**: Making our own choices and authentically expressing ourselves, which suffers in the face of triggers like coercion, loss, and unfairness (reference page 103).
  - **Sample shadow goals**: Dominance, rebellion, giving up (see the second table at the end of this chapter for a list).
  - **Sample shadow habits**: Micromanagement, rejecting authority, silence (see the second table at the end of this chapter).
- **Choice-crafting tools**:
  - **Bully jujitsu**: Using your opponent's strength against them. Example: humor destabilizes the enemy, neutralizes fear, and fosters the courage to act.
  - **Choice support**: Offering options and saying, "At the end of the day, it's up to you."
  - **Choice-supportive leadership**: Validating others' experiences, offering information for authentic decisions, and supporting empowered action.
  - **2-2-2 tool**: When you're ready to give up, ask, "What do I need in the next two minutes, two hours, and two days?" Give yourself a break, then get back at it!
- **Moving forward from loss**:
  - **Choose one thing**: Even small acts of need crafting in one domain of our lives improve need satisfaction in others. Ask: "What is one thing I *can* control?"
- **Trading mustivation for motivation**:
  - **Authenticity check**: "How do I *really* feel about doing this?"
  - **Avoid overidentification:** Escape the experience where one part of yourself becomes central to your identity, like work, by building a balanced identity.

- **Build a balanced identity**: 1. Separate your role from your identity ("What I do is not who I am"). 2. Set limits (recognize your true capacity for external pressure). 3. Create an identity hierarchy with a clear order (e.g., family first). 4. Try on a temporary role (escape into an entirely different identity—e.g., poker player).
- Scientifically supported need-crafting leisure activities:

Confidence-Supporting Activities	Choice-Supporting Activities	Connection-Supporting Activities
• Acting • Arcade games • Baseball • Bicycling • Chess • Computer games • Deep-sea fishing • Poker • Pool • Soccer • Sports clubs • Tennis • Weightlifting	• Acting • Baking and cooking • Camping • Ceramics • Collecting autographs • Dancing • Drawing • Gardening • Guitar • Hiking • Needlepoint • Painting • Photography • Quilting • Volunteering: Crisis intervention • Volunteering: Scouting • Woodworking	• Acting • Baseball • Church and religious group meetings • Dancing • Frisbee • Live music • Picnicking • Soccer • Socializing • Sports clubs • Visiting friends and relatives • Volunteering: Medical setting

Thwarted Choice: Shadow Goals and Habit

Category	Shadow Goals	Sample Shadow Habits
Protect: Fighting back to avoid guilt, shame, or blame.	• **Rebellion**: A drive to blindly defy expectations or rules.	• Pushing our/others' boundaries. • Taking risks regardless of the consequences. • Refusing to follow social norms. • Rejecting authority through bold disobedience. • Doing the opposite of what's expected (the boomerang effect). • Taking dangerous risks. • Passive aggressiveness (e.g., intentional stalling). • Lying or cheating.

Shatterproof Transformations and Tools

Category	Shadow Goals	Sample Shadow Habits
Prove: Seeking external evidence about ourselves or our place in the world.	• **Dominance**: A drive to exercise control over others. • **Restriction**: A drive to control or limit one's own choices.	• Making unilateral decisions. • Restricting others' input or choice. • Refusing to delegate. • Using coercion to control others. • Extreme micromanagement. • Hostility, aggression, or bullying. • Extreme pursuit of order within ourselves or our surroundings (e.g., diet, exercise, or other habits).
Prevent: Denying experiences, or devaluing our needs, to feel better right now.	• **Harmonizing**: A drive to do what we "ought" to do to make things "easier." • **Giving Up**: A drive to surrender without resistance.	• Allowing apathy or amotivation to set in. • Silence, passivity, or disengagement. • Ignoring or denying the existence of constraints. • Adopting beliefs, emotions, behaviors, or values that don't reflect our authentic selves. • Presenting a facade of compliance to avoid rocking the boat. • Suppressing our real feelings. • Denying our needs, comfort, or happiness. • Self-medication (food, alcohol, gaming). • Doing what others want, even though we don't want to. • Believing we're powerless to change our circumstances. • Supporting controlling institutions.

Chapter 11

Crafting Connection

> "We are like islands in the sea, separate on
> the surface but connected in the deep."
> —William James

Charlotte was living in paradise. As she strolled down the cobblestoned streets of Amsterdam, the world she'd left behind in Ohio felt so flat and hard and overly bright.

The house that Charlotte, her husband, and their young daughter now inhabited was a gorgeous, high-ceilinged duplex with tall windows looking out on the water across city roofs and the glowing spire of the Oude Kerk. On Saturday mornings, she and her daughter would treat themselves to Dutch pancakes smothered in bacon and cheese, a local delicacy to which they'd taken an obsessive liking. Aside from the occasional near miss with one of the city's many cyclists, Charlotte had nothing to complain about. Nothing whatsoever.

And yet. Why couldn't she feel *happy*?

Charlotte could see the beauty, history, and culture all around her. She could taste the delicious food. She could hear her daughter's carefree laughter. She could understand, at least intellectually, that they'd moved to one of the world's most perfect cities. But somehow, none of it managed to soak past her skin. It was like watching herself in a movie: a colorless, lifeless, monotonous movie.

But since Charlotte had nothing to complain about, she wouldn't. She would focus on being an attentive mother and agreeable wife to her hardworking husband, Ian. Life was good enough, which was much more than millions of people with *real* problems could say. And she should feel grateful that Ian's job allowed her to embark on such an amazing adventure, even though it meant her career had to take a back seat for a while, and his demanding work at a global bank took him away for weeks at a time.

And boy, those weeks got long and lonely, especially during the city's dark, cold winters. It wasn't easy to make friends in Amsterdam. The Dutch were polite and kind, but in a detached way that would be seen as outright rude in small-town Ohio. Charlotte appreciated their intolerance for nonsense, but feared her effusive midwestern demeanor was something of a barrier to connection.

Every evening while her daughter studied, Charlotte curled up on the sofa with a blanket over her legs and Netflix on her laptop, trying to tune out the sound of the drunken revelry on the icy canals outside her window. Some nights, when she felt especially down, she'd relive fond memories in her journal.[1]

Thankfully, Charlotte also had her regular FaceTime calls with her best friend to look forward to. Zoe's face appearing on her iPad felt like the sun coming out. The pair had always been close, but in moving across the world, Charlotte had come to appreciate that old saying about absence and the heart growing fonder. It really was so true!

Charlotte's calls with Zoe were the highlight of her week, especially when Ian was gone, which was a lot. On what she came to call "Zoe days," she'd hop out of bed and then spend the day bouncing around until they spoke. Afterward, she'd replay their conversations, smiling at all the funny, cute, incisive things her friend had said.

On a snowy February day, Charlotte confessed her loneliness to Zoe, who shared that she, too, had been struggling and that her marriage was in trouble. Was it wrong that this made Charlotte perk up? Of course, she

---

1. Charlotte was on to something: Nostalgia is indeed one research-backed tool to help us temporarily compensate when we're feeling lonely or disconnected (Johannes Seehusen et al., "Individual Differences in Nostalgia Proneness: The Integrating Role of the Need to Belong," *Personality and Individual Differences* 55, no. 8 [2013]: 904–908).

didn't want Zoe to be unhappy. But the hushed, intimate conversation and tears of sadness and relief brought them even closer. Soon, they were FaceTiming several times a week.

One night, when she called Ian at his hotel in London, he snapped at her, "Jesus, Charlotte, all you talk about is Zoe. It's like you've got a schoolgirl crush on her."

Suddenly, it all clicked. Hanging up the phone, Charlotte flashed back to a memory of a childhood family vacation. They had arrived in Oregon after dark and gone to sleep; in the morning, she opened the curtains and was awestruck to see, through the hotel window, the magnificent view of Mount Hood framed against a perfectly clear blue sky. That was how this revelation of her love for Zoe felt—as sudden and massive and undeniable as that mountain.

Charlotte was completely unprepared for this twist. Her whole life, she'd proudly identified as a "good girl." Straight As in school? Check. A star athlete at a top university? Check. A good career, but still pretty and nonthreatening? Check, check, and check. Rather than try to figure out who she was and what she truly needed, it had always been easier to just be who people wanted her to be: low-maintenance and endlessly accommodating. With her husband in particular, she tried to be the "cool wife" who watched sports and served home-cooked meals and never complained.

So . . . falling in love with a woman, while married to a man? Crazy. Nuts. *No effing way.*

Charlotte commenced an avoidance strategy. Maybe Ian was right—this *was* just a silly crush. If she could just ignore her feelings, surely they would eventually fade away.

She pared back her calls with Zoe. She messaged her less. She tried to force her life back into the shape it was before this craziness came over her. The adjustments were painful, and she knew they hurt Zoe too. But good enough would have to be good enough. What other choice did she have?

This strategy began to unravel when two friends—Natalie and her husband, Brian—visited Amsterdam. As the three old schoolmates spent a spring morning biking through the city's picturesque streets, the couple asked Charlotte about her favorite places, what inspired her, and what she loved most about her new city. For the first time in a year, outside of her calls with Zoe, Charlotte felt fully *seen*—like Zoe, Brian and Natalie cared about what she had to say. They wanted to make sure she was happy. They loved her for who she was rather than who they wanted her to be.

That day, Charlotte felt filled up, not lonely for a second. This gave way to a "tsunami of realization" about the depth of the disconnection she'd been feeling. It had nothing to do with being transplanted into a strange city. It wasn't about Dutch aloofness. It wasn't even about Ian's travel. (She felt just as alone when he was home.) The problem was that her marriage was no longer fulfilling the most fundamental of all human needs.

CONNECTING IN AN AGE OF DISCONNECTION

At our core, humans are relational beings. In the beginning, we lived in small, stable groups of about 100 to 150 people, and sticking together was a core survival strategy. By sharing labor, resources, and information, and cooperating to hunt and fend off threats, early humans could survive and ultimately reproduce. (This may also explain why we especially crave the care and support of others when we're feeling sick, scared, or in danger.) To survive, humans had to cooperate[2]—and to cooperate, they had to care about one another. And so, we are wired to avoid loneliness and crave love and acceptance. The level of genuine connection we experience in our lives depends on where we fall on this spectrum:

Loneliness ⟷ Love

---

[2]. Though human brains instinctively default to altruism, selfish tendencies may have evolved later due to resource scarcity (Jennifer Crocker, Amy Canevello, and Ashley A. Brown, "Social Motivation: Costs and Benefits of Selfishness and Otherishness," *Annual Review of Psychology* 68 [2017]: 299–325; Caroline Roux, Kelly Goldsmith, and Andrea Bonezzi, "On the Psychology of Scarcity: When Reminders of Resource Scarcity Promote Selfish (and Generous) Behavior," *Journal of Consumer Research* 42, no. 4 [2015]: 615–31).

Research reveals two basic building blocks of connection. The first is **belonging**. Because social bonds help us work together, we're built to form them easily. For instance, one study showed that upon receiving a holiday card from a complete stranger, most people reciprocated by sending one back. Bonding becomes especially effortless in adversity; as any military veteran will tell you, lifelong ties are "forged under fire." And we've all experienced how everyday moments of belonging—a cup of coffee with a like-minded colleague, the camaraderie of an intramural volleyball team, the higher power of a religious service—buffer stress and make us feel seen, heard, appreciated, supported, surrounded, championed, and defended.

The second connection building block is **relationship depth.** Because our ancestors spent their lives among the same relatives and other close contacts, these relationships had to be strong enough to stand the test of time. Relationship depth is fostered through interactions that boost trust and intimacy, leaving us with the sense that others truly care for us—and especially under stress, that we can rely on them for love and support. Our deepest relationships (with romantic partners, best friends, family, or even colleagues, bosses, and clients) function reciprocally, meaning that we both give *and* get care and support. We feel less connected when others take more from us than they give, and when others give more to us than we give them.

Belonging and relationship depth remind us that we're not alone, which boosts our sense of self-worth and dignity.[3] This doesn't just feel good; it literally rewires our brains. Close connections trigger the release of feel-good chemicals like exogenous opioids. They've been shown to improve cancer survival rates (even after factoring in differences in income and medical treatment) and protect against death over the course of a decade. On the flip side, loneliness is as bad for our health as smoking fifteen cigarettes a day, increasing our risks for dementia and heart attack. It's also strongly linked to mental health challenges and may be a

---

3. In the introduction, I mentioned one of composer Stephen Sondheim's most famous lines from *Into the Woods*: "You are not alone / No one is alone." I've always wondered why that line hits you in the gut, and now I understand: Our need to *not* be alone is primal and visceral.

contributing factor to the rise in mass shootings. According to former US surgeon general Vivek Murthy, this loneliness epidemic has become an urgent public health concern.

How did we get here? As I mentioned earlier, our ancestors generally lived in the same place with the same people for their entire lives, making it easier to deepen connections over time. Nowadays, we frequently uproot our lives and start over in new places, like Charlotte did. While our forebears lived in "extended kin networks," with uncles, aunts, cousins, and grandparents, many cultures today live in nuclear families and have fewer extended family interactions. And where early humans would select from one or two dozen potential mates, we swipe through an endless number of complete strangers we'll never even meet.

It shouldn't come as a surprise that **social connections have been sharply declining for decades.** Across gender, age, and ethnicity, in-person socializing has dropped 30 percent over the last fifty years. We aren't just forging fewer close friendships; we're spending twenty fewer hours on average with friends per month than we were twenty years ago. Demographic trends like falling family size and marriage rates further hinder close connections. As a result, only 39 percent of US adults feel "very connected," and nearly half experience persistent loneliness. These effects are especially pronounced for young people, who are almost twice as likely to feel isolated. In one heartbreaking survey, almost one-quarter reported having *zero* friends.

As Derek Thompson writes in the *Atlantic*, "There is no statistical record of any other period in US history when people have spent more time on their own." Of course, technology has played a key role, monopolizing our attention, displacing in-person interaction, and reducing relationship quality. (Everyone has that friend who can't put their phone away at dinner—or maybe that friend is you!) And while the percentage of teens who are online "almost constantly" has doubled since 2015, research shows that it takes just *one hour* of daily screen time (on average) to cause loneliness, depression, and anxiety.

But technology isn't entirely to blame. On a broader societal level, we're experiencing what Derek Thompson dubs "a kind of ritual recession" stemming from sharp declines in community-based routines and more options for sealing ourselves off from others. Participation in collective activities—book clubs, sports teams, communal worship—has dropped by more than one-third in the last twenty years, and fewer than two in ten Americans feel "very attached" to their local community. According to sociologist Robert Putnam, trust in social institutions and other people has reached near-historic lows. Finally, the shift to virtual and hybrid work has left the workplace—arguably the last intact bastion of community—increasingly fractured.

STARVED FOR CONNECTION

From small snubs to big betrayals, connection triggers can cut deep, leaving us feeling abandoned, alone, unappreciated, unsupported, jealous, furious, and betrayed.

One major trigger is **rejection.** Whether it's not being invited to a friend's wedding, being passed over for a job, getting ghosted by a date, or receiving the silent treatment from your partner, rejection is the best-established cause of anxiety and hostility. It damages our self-control, reasoning, and even IQ. And since being excluded from the tribe threatened our ancestors' very survival, human brains are wired to respond to rejection—even the slightest snub can set off our rejection alarm. Studies show that merely watching someone else being left out stirs up negative feelings, as if it's happening to us.

Another disconnection trigger is **neglect,** which happens when important people or groups fail to show us care or concern. This can take the form of cold and distant interactions, the absence of support, or seeing others getting preferential treatment. One especially sneaky form of neglect, known as conditional regard, happens when others only show affection or approval when we behave the way they want, essentially forcing us to choose between satisfying our choice or our connection needs.

A third trigger is **conflict,** which can take the form of interpersonal tension, disagreements, and misunderstandings—like a family feud, a power struggle at work, or a dispute with a friend. Conflict can stem from differences in styles and values, poor communication, competing priorities, and occasionally, harmful intentions. It is painful not only because of the friction it creates with others but also because it can threaten our personal interests, goals, or status. At a deeper level, conflict challenges our core desire to be liked and accepted.

The fourth trigger, **cruelty,** describes dehumanizing or hurtful behavior, whether intentional or not: abuse, attacks, punishment, manipulation, humiliation, being exploited for someone else's gain, etc. Whether delivered overtly or through "subtler" microaggressions, cruelty threatens our dignity and self-worth—which can fuel deep feelings of shame. One all too common form of cruelty is, of course, bullying; a whopping 84 percent of people report having been bullied, according to one study. And contrary to the popular belief that bullies prey on the weak, most tend to target socially connected superstars, especially women in higher-status positions.[4]

Finally, **betrayal** occurs when someone whom we feel is "safe" breaks our trust, loyalty, or fidelity. Whether it's a friend spilling a secret, getting ghosted by your best friend at work, or a partner being unfaithful, betrayal doesn't just shatter our sense of safety; it signals that someone we find important doesn't value us the same way, making us feel like our entire relationship (or, depending on the relationship, *life*) was a lie. Typically, the more someone's behavior violates our expectations, the bigger fire it ignites within us. Humans are wired to spot betrayal, fueled by what researcher Glenn Geher and colleagues call the "suite of cheater-detection adaptations" that helped our ancestors avoid exploitation. In the ancient past, the ability to distinguish friend from foe was a matter of life and death, so it was better to stay furious than risk being betrayed again.

This underscores an important truth about connection triggers: Even small slights—like a friend forgetting your birthday—can be disproportionately triggering. To see why, consider that at any given moment, we are bombarded with literally eleven million pieces of incoming information.

---

4. Make sure to check out my friend Amy Cuddy's brilliant book *Bullies, Bystanders, and Bravehearts* for more data and practical tools for dealing with bullies through the lens of her own (absolutely horrific) experience.

To cope, our brains default to autopilot, making swift, automatic judgments about our environment—including other peoples' motives and behaviors. While often helpful, this system falters under stress, distorting our perceptions and skewing reality. And because false positives were much less risky for our ancestors than false alarms, humans are wired to over detect interpersonal threats from others. Let's say you're on the couch scrolling through social media and see that your spouse, who you assumed was working late, is out to happy hour with a few coworkers. The same social threat detection system that helped early humans notice a mate's infidelity might trigger feelings of righteous indignation or moral outrage, even if no actual betrayal occurred.

And the shadow goals these visceral responses create are just as varied as their triggers. First, there's the drive to prove our importance to others by chasing things like **popularity** (proof of being liked) and **validation** (proof of being valued). Alternatively, to protect ourselves from shame, we might resort to **spite** (hurting those who've hurt us) or **aggression** (lashing out to prove our dominance). Finally, to prevent pain, we might choose **seclusion** (closing ourselves off) or **pretending** (avoiding rocking the boat). (For a list of sample common habits for connection threats, see page 202.)

Returning to Charlotte's predicament, we now see that she was grappling with a few distinct triggers: neglect from her husband's frequent work travels, the cruelty of his indifference when he was home, and a dash of rejection from several unsuccessful attempts to make friends in Amsterdam. In response, "good girl" Charlotte opted for the shadow goals of pretending and seclusion, adopting shadow habits like denying the true toll of her disconnection—even distancing herself from Zoe.

Such severing is significant, even when we have other relationships to lean on. That's because research shows that our single deepest connection (in Charlotte's case, Zoe) has the biggest impact on our mental health and well-being, with additional attachments yielding diminishing returns. These relationships, dubbed **pair bonds** by researchers—and "my person" by the cast of *Grey's Anatomy*—are often romantic, but they don't have to be. Either way, pair bonds provide a psychologically safe base from which to live our lives. The worst thing we can do in the face of frustrated connection, then, is to push our pair bonds away, as Charlotte instinctively did.

Luckily, something in her shifted after that day with Brian and Natalie. Charlotte's "tsunami of realization" about the state of her marriage and true feelings for Zoe was a sentinel event, jolting her out of complacency and helping her finally see what she'd been missing: that the opportunity for a life filled with deep love and connection was hers for the taking. All she needed to do was find the courage to pivot.

BACKERS AND BARNACLES

I learned the term "backers and barnacles" from my client Paul and his wife, Courtney, the cofounders of a fast-growing construction company. Paul had navigated more than his fair share of adversity. Shortly after founding the company, he suffered a devastating accident that left him unable to walk and hospitalized with a severe staph infection. He recovered, but as if that setback wasn't enough, after growing the company for over fifteen years, he was then forced to oversee its bankruptcy after a failed acquisition.

Through these trials, Paul found, the people in his life unwittingly sorted themselves into two buckets, and he could have never predicted who would end up where. **Backers** showed up and stuck by his side, no questions asked. Like the engine on a motorboat, they helped Paul propel himself through turbulent waters. Then there were the **barnacles**: people who hung around during the easy times but were unwilling or unable to show up for him when things got tough. Like marine crustaceans clinging to the hulls of ships, barnacles drag us down and get in the way of forward movement. By leaning heavily on his backers—and scraping his barnacles loose—Paul managed to navigate both crises in true shatterproof fashion. First, he defied his doctor's predictions by taking ten thousand steps at Disneyland with his family. Later, the company's growth was setting industry records.

But here's the tricky thing: when times are good, backers and barnacles can be virtually indistinguishable. This is why hard times are an excellent, albeit painful, litmus test for our relationships. I learned this lesson firsthand when, after being diagnosed with the EDS trifecta, more tests revealed I'd become alarmingly immunocompromised. Amid a global pandemic, this created a simple litmus test. When a friend or family member texted me wanting to get together, I'd explain my situation and ask if they'd be willing to meet me outdoors, take a COVID test, or wear a mask because my immune system wasn't working. To my great dismay, about half the time, the text chain would suddenly go dead—and, as Paul warned me, often with the people I least expected. Ultimately, the reason didn't matter. Anyone willing to let a relatively minor personal inconvenience trump a simple action to minimize the very real risk of, you know, *my untimely death*? Clearly not one of my people.

And so, for about eighteen months before I began immunotherapy, I learned a harsh but important truth: when life's struggles transform you from a bubbly, convivial person into a vulnerable, broken human who desperately needs someone to show up for you, not everyone will. Yet we might see the surprise and pain of unmasking barnacles as a short-term tax in service of deepening our greatest connections. If someone frustrates your needs more than fulfills them, it's a clear sign you've got a barnacle on your hands.[5] And trust me, **it's a gift to definitively find out who "your people" are and who they aren't.**

CRAFTING TRUE CONNECTION
(BY DISCONNECTING FROM THE PAST)

Once we find the courage to take a good, honest look at our relationships, we may discover that the people we thought we belonged with are no longer providing the support, commitment, or intimacy we need. This can be terrifying, and Charlotte was no exception.

Still, she vowed to build a future where she'd be surrounded by people who loved and supported her. Wanting to visualize exactly what this

---

5. I would like to thank the astute beta reader who suggested this idea. It is absolutely brilliant and rings deeply true for me personally.

connected life could look like, Charlotte crafted a vision board—a mosaic of her hopes and dreams for a better future.[6] At this point, you probably have a pretty good idea of what it included: photos of her and her daughter laughing, Zoe and her kids, a screenshot from a FaceTime where Zoe was looking at Charlotte with "so much love," a random image she pulled from Google of a quaint street where she could see herself living, a photo of a big front porch with a swing. Yet in Charlotte's newly envisioned life, one face was conspicuously absent. It was at this point that she finally gave in to the voice in her head. *Ian isn't my person, and that's okay.*

The next part was tough. Even though she knew she was doing the right thing, Charlotte felt immense guilt for upending her marriage, even after two years of unsuccessful couples counseling. It didn't help that she was met with disappointment, resistance, and judgment from her work colleagues, her closest friends, and even her family. And of course, she had devastated Ian; when she asked him for a divorce, he initially thought she was kidding and then grew hurt and angry. Charlotte was especially heartbroken to see his distress at the brutal realization that he wouldn't live with his daughter full time.

By walking away from her marriage, Charlotte was—as the great Brené Brown would say—standing squarely in her own wilderness. But while the whole world seemed to be telling her that what she was doing was wrong, she clung to her vow: *I will never again allow good enough to be good enough.*[7] Before long, she discovered that whenever negative judgments

---

6. While the science on vision boards is scant, research shows that visualization, when paired with action, leads to better outcomes than action alone, and visualization is most effective when we imagine the *process* rather than just the *outcome* (Hwi-young Cho, June-sun Kim, and Gyu-Chang Lee, "Effects of Motor Imagery Training on Balance and Gait Abilities in Post-Stroke Patients: A Randomized Controlled Trial," *Clinical Rehabilitation* 27, no. 8 [2013]: 675–80; Lien B. Pham and Shelley E. Taylor, "From Thought to Action: Effects of Process-Versus Outcome-Based Mental Simulations on Performance," *Personality and Social Psychology Bulletin* 25, no. 2 [1999]: 250–60).

7. If you're conflicted about Charlotte's decision, research shows that while divorce is difficult, staying together can be worse for couples and their children. For couples, staying in unhappy marriages is associated with poorer physical and mental health than divorce (Daniel N. Hawkins and Alan Booth, "Unhappily Ever After: Effects of Long-Term, Low-Quality Marriages on Well-Being," *Social Forces* 84, no. 1 [2005]: 451–71). For children of unhappy marriages marked by conflict—including silence, withdrawal, and emotional unavailability—their long-term well-being

popped up in her mind, they were often founded in fear. "Sometimes," Charlotte explains, "it took a backhoe to find the nugget, but I usually got there." Importantly, she learned to forgive herself for the inevitable consequences of pursuing what she knew she needed in her life.

Today, Charlotte and her daughter reside in her hometown of Kenwood, Ohio, with Zoe and Zoe's two kids. Blending their families was anything but easy—in fact, it was the "hardest, hardest thing"—but they made it through, having approached the challenge with grace, kindness, and love. After "a lot of work," Charlotte has built a harmonious relationship with the man she now lovingly refers to as her "wasbund." Their daughter is thriving, remaining close to both her parents. She enjoys traveling the world to visit Ian abroad while cherishing her new family unit in Ohio. Oh, and remember the random street from Google Images Charlotte pinned to her vision board? As destiny would have it, she now happens to live on that *exact street* (!) and delights in sitting on her big front porch swing with her partner and their three beautiful kids.

Charlotte has never been happier. But while her life is now a dream come true, she's careful to point out that it isn't perfect. Because, she has learned, nothing is. Despite all the ways in which her life has changed, the biggest difference now is her mindset. Instead of forcing herself to give in to what was expected of her, she pursues what she truly wants and needs while holding compassion for herself and others.

The moral of this story is that **no matter what anyone tells you, it's okay to want more out of your life and to go after it, even if it means you have to walk away from something.** And because huge changes often bring huge rewards, the most powerful transformations often begin with the decision that in service of true connection, big risks are worth taking.

---

is often worse as well (Patrick F. Fagan and Aaron Churchill, "The Effects of Divorce on Children," *Marri Research* 1, no. 1 [2012]: 1–48; Gordon T. Harold and Ruth Sellers, "Annual Research Review: Interparental Conflict and Youth Psychopathology: An Evidence Review and Practice Focused Update," *Journal of Child Psychology and Psychiatry* 59, no. 4 [2018]: 374–402; Kelly Musick and Ann Meier, "Are Both Parents Always Better Than One? Parental Conflict and Young Adult Well-Being," *Social Science Research* 39, no. 5 [2010]: 814–30). As a child of divorce myself, I can attest that it is indeed awful, but what would have been *way* worse was watching my parents stay married!

OTHER WAYS TO CRAFT CONNECTION

Despite everything she went through, Charlotte was lucky that she found "her person" in someone who had been in her life all along. But what happens when we look around us and the support, belonging, or intimacy we crave is nowhere to be found?

Turning Conflict into Connection

Charlie, who worked in sales at a software company, had always been someone who could get along with anybody. So when a new boss parachuted in, he greeted the change with open arms. Sure, Hank was green and young enough to be one of his kids. And maybe the new manager's communication skills left a little something to be desired. But Charlie remained confident they could figure everything out.

Take, for example, their first performance review. While it hadn't been the *most* pleasant exchange, Charlie says it "certainly wasn't the worst ever." After all, Hank had pointed out a few truly helpful development opportunities, albeit clumsily.

One evening a few days later, an email marked "high importance" landed in Charlie's inbox with the ominous subject line, "Your Slack Note." Initially, Charlie was confused. What was this about? He opened the email to find a lengthy rant from Hank chastising him for "airing your dirty laundry" on Slack. Then he remembered. Earlier that day, while surfing a company Slack channel—one that Hank wasn't even *on*—Charlie stumbled upon a colleague's post about an excruciating gym session following a long hiatus, saying she'd never sweat so much in her life. Charlie had playfully joked, "Well, then, you've probably never had a really bad performance review!"

Someone, it seemed, had shared this exchange with Hank. And Hank had clearly assumed Charlie's lighthearted comment referred to the performance review they'd just had and that it was a jab at his management style. Heart racing, Charlie hastily composed a response reassuring Hank that he absolutely hadn't been talking about their review. *This is just a misunderstanding.* Charlie thought, *I'm sure it will be fine.*

But it was not fine.

When Hank flat-out refused to believe Charlie's explanation, a surge of red-hot rage rushed through his veins. He had never been anything but honest with his boss. How *dare* Hank refuse to take him at his word? "I was always good at vitriol," Charlie concedes, "for better or for worse."

Indeed, most people are. It may not surprise you that **the biggest source of daily anger is other people.**[8] The rage and indignation of being misunderstood is anger in its rawest form, especially when our attempts to correct the record or the wrong fall flat, as Charlie's did. Yet while no one enjoys anger, inherently, it's neither good nor bad; it's simply a hardwired response to a threatened need. On the plus side, it can push us to stand up for ourselves, like when Charlie responded to Hank's email to set the record straight. But once we let anger morph into aggression or spite, it usually backfires—even when lashing out feels good in the moment. In Charlie's case, threatening or attacking Hank would have been (as an old boss of mine says) a CLM: career-limiting move!

Yet while aggression and spite can be unhealthy responses to anger, so is bottling it up. This can lead to a shadow habit known as **offense rumination,** where we nurse our fury in solitude, replay the injustice over and over, and concoct elaborate revenge fantasies. Offense rumination hampers conflict resolution, breeds paranoia and anger, and creates cognitive chaos (in Charlie's case, that Hank was a manipulative monster intent on ruining his career).

I once stumbled upon an interview with a former counterterrorism officer who offered a profound observation about human nature. "The one thing I learned in the Agency," she explained, "is that everyone thinks they're the good guy." What's more, when we feel wronged, our sense of self-importance and entitlement skyrocket. It's easy to cast

---

8. When Jean Paul Sartre argued that hell is other people, he was on to something!

ourselves as the righteous hero, convinced of the purity of our motives, all while we demonize those who've wronged us and scrutinize their actions for further signs of evil. I call this **bad guys bias,** and it's a sneaky habit fueling offense rumination. Bad guys bias doesn't just burn precious mental resources; it sparks aggressive and socially inappropriate behavior that further intensifies our feuds. This is precisely why well-intentioned mix-ups can metastasize into full-blown wars.

After weeks of ruminating about Hank's unfair accusation, Charlie decided to take a calculated leap of faith. Recognizing that conflict is rarely one-sided, he conceded that maybe he was playing a role too, even if he couldn't see it yet. (After all, to paraphrase the comedian Ricky Gervais, just because we're offended doesn't mean we're right!)

Luckily, as Charlie discovered, breaking free from bad guys bias and offense rumination might be simpler than we think. (Incidentally, if you're curious about your tendencies for both, check out Appendix J.) As it turns out, the key to breaking these habits lies in the brain's **exploration network.**[9] When this network is activated, we're more likely to generate creative, out-of-the box ideas we wouldn't normally consider; when it's deactivated, we fall back on old ideas and cognitive shortcuts.

One trick to activate our exploration network is to get curious about our situation with someone we trust. Charlie decided to swap his enraged "one-sided bitch sessions" with his wife for more "exploratory dialogue." He noticed this shift essentially boiled down to one question: **What if I'm wrong?** This challenged Charlie to consider other explanations for Hank's behavior—ones where he *wasn't* an evil villain. Perhaps Hank's email stemmed from his insecurity about giving performance reviews. Maybe Hank was resentful about a tough appraisal he'd received from *his* boss and was projecting those feelings onto Charlie. Or maybe he'd just had a bad day.

---

9. If you read *Insight*, this is the same network I refer to as "default mode," which is activated when we practice mindfulness. More recent research has shown that it serves other functions besides rest (Michael Platt et al., "Perspective Taking: A Brain Hack That Can Help You Make Better Decisions," *Knowledge at Wharton*, March 22, 2021, https://knowledge.wharton.upenn.edu/article/perspective-taking-brain-hack-can-help-make-better-decisions).

In addition, Charlie embraced the concept of "the crazier, the better," letting his imagination run wild. He brainstormed less likely but more amusing explanations: maybe Hank had sent the email in an exhaustion-induced fugue state; maybe he was a method actor preparing for a role as an emotionally reactive boss; or perhaps Hank's email was a coded message from an alien civilization! The goal of such **creative perspective taking** isn't to find the answer but to activate our exploration network—as a bonus, Charlie found that his wild speculations also took the edge off his anger.

In sum total, these actions helped Charlie see that Hank wasn't trying to ruin his life. Hank was just a young, inexperienced manager who'd probably fired off a late-night email without much thought and then dug into his position to save face (just like Charlie had dug into his). Plus, forcing himself to consider his boss's perspective helped Charlie better understand how his casual Slack comment might have landed. He *had* equated performance reviews with a particularly excruciating gym session, just days after his own.

Already feeling more empowered and less paranoid, Charlie knew that he needed to own his part in the mess. So, he set up a Zoom meeting to address the issue and apologize to Hank. The conversation went fairly well, and the pair eventually found some—not a lot, but some—common ground.

That Zoom was a turning point for Charlie. Now, anytime he feels his temper rising because of something Hank says or does, he remembers that "neither one of us is trying to make the other unhappy." To this day, even though Hank's communication style hasn't changed one bit,[10] Charlie no longer sees him as an enemy but someone grappling with his own insecurities and emotions. They still butt heads every now and then, yet this new understanding has deepened their relationship, allowing them to work out almost any issue with less drama. More broadly, Charlie can confidently distinguish between "temporary intense aggravations"

---

10. You may notice that this is a theme in shatterproof transformations—the simple fact that we often can't change other people is another reason transforming ourselves is so vital.

and situations where something bigger is going on. It's a journey, but as Charlie put it, "It's absolutely worth it."

Indeed, when we dig into our knee-jerk reactions, we usually discover that turning conflict into connection becomes easier than we think. When in doubt, we should remember one thing: To paraphrase British musician Dave Mason, there aren't good guys or bad guys; there's just you, and there's just me, and we just happen to disagree.[11]

Something Greater Than Ourselves

To anyone who knew her, Helen's life seemed ripped from the pages of a glossy magazine—a thriving business, fabulous clients, a wonderful marriage, and great friends. Yet deep down, she felt a gnawing emptiness. She was finally running the company she'd always dreamed of, but instead of feeling excited, she felt flat, demotivated, and disconnected. Initially, the entrepreneur attributed this malaise to stress. If she could just manage to set aside a bit more "me time," maybe her spirit would feel more grounded. But even after she started practicing yoga every morning and worked regular swimming sessions into her busy schedule, these feelings persisted.

One balmy summer afternoon, Helen found herself seated beside her husband in a beautiful park in Priego de Córdoba, Spain, where they'd been living for three months. Lost in thought, she glanced over her husband's shoulder, where his Kindle illuminated a familiar passage from a book by Christian author Tim Keller: "The thing is, when you put anything in the place of God, that is not God, you will only ever be disappointed." In *Counterfeit Gods*, Keller explains that everyone, knowingly or not, makes sacrifices to some kind of god in the name of finding fulfillment and connection. Take, for instance, the couple who so desperately

---

11. As a final point, I do not want to imply that we are all "good guys," all of the time. There are certain factors that prevent people from, for example, feeling empathy or controlling their violent urges. My point here is that in most situations, even if we think we're dealing with an evil mastermind, it's also possible (likely?) that they are not one.

want to start a family that they sacrifice time, money, and mental health to go through rounds and rounds of IVF, desperate for a baby that finally will complete them.

Prior to this point, Helen's decade-long journey as a Christian was filled with ups and downs, leaving her convinced that she didn't need God in her life. But that day in the park, the words on her husband's Kindle hit her like a lightning bolt. She instantly recognized the shadow goals she had been chasing—wealth, free time, the perfect body (hence the obsession with yoga and swimming)—were counterfeit gods that ultimately wouldn't give her what she *truly* needed: a connection to something greater than herself. Thus, Helen's spiritual journey began again.

The shatterproof goal of **spirituality** needn't be about God like it was for Helen, though it certainly can be. It simply means discovering and preserving the sacred in our everyday lives, however that looks to us. We can find spirituality in religion, nature, meditation, creative expression, service, community, or anything else that helps us feel connected to the world. Research underscores the transformative power of spirituality, provided it's intrinsically (versus extrinsically) motivated. This deep, authentic pursuit of the sacred powerfully predicts wellbeing, even without traditional practices like attending religious services. At the core of spirituality lies **awe,** an experience where the ordinary becomes extraordinary, prompting us to see the world as vast and transcendent. By shrinking the self and broadening our perspective, moments of awe foster a deep sense of connection to a force greater than ourselves.

In this way, spirituality offers another pathway to connection: one Helen was determined to take. Encouraged by her husband and a close friend, she embarked on the journey of realigning her habits with her shatterproof goal—forsaking counterfeit shadows and focusing on her faith. Though doubts occasionally crept in, Helen found solace in deepening her understanding of Christianity and forging a stronger connection with something bigger than any one person or relationship. With each day, she embraced the practice of placing God at the center of her life. And while it wasn't easy, Helen gradually felt the burden of false expectations ease away—finding liberation in understanding that salvation stems not from striving, but from faith and love.

Ten months later, Helen's life looks the same on the outside. But on the inside, everything is different because *she* is different. Now, her rediscovered spirituality is infused in every aspect of her life—a life that brims with joy, peace, and the freedom of feeling like herself again. All her chains are gone, as she puts it, and knowing her commitment is here to stay helps her confidently face the future.

That's the thing about crafting connection. It lifts us beyond the mundane, awakening the feeling that we matter in this world and *to* this world. In the process, we lift up the people around us as well. As existentialist author Albert Camus once put it, "When you have once seen the glow of happiness on the face of a beloved person, you know that a man can have no vocation but to awaken that light on the faces surrounding him."[12]

CHAPTER 11: KEY TAKEAWAYS AND TOOLS

- **Love versus loneliness**: Humans are wired to avoid loneliness and crave love.
- **Connection triggers**: Connection suffers in the face of triggers like rejection, cruelty, and betrayal.
  - **Sample connection shadow goals**: Popularity, aggression, seclusion (see the table at the end of this chapter).
  - **Sample connection shadow habits**: Bad guys bias, offense rumination, paranoia (see the table at the end of this chapter).

---

12. Just in case you're noticing the number of existentialist themes and references in this book, this is the school of thought (or rather, antischool of thought) with which I have always most closely identified.

- **Building blocks of connection**: (1) belonging (frequent, positive interactions), (2) relationship depth (trust and intimacy-boosting interactions, reciprocal relationships).
  - **Pair bonds**: Our single closest and most important connection ("our person").
  - **Nostalgia**: A tool to temporarily compensate for disconnection by recalling past moments of connection.
- **Finding "our people"**: Build your life around your backers; everyone else is just window dressing.
  - **Backers**: People who show up and stick by our side, no questions asked.
  - **Barnacles**: People who are unwilling or unable to show up when we need them most.
  - **Relationship litmus test**: When times are good, backers and barnacles are often indistinguishable. The next time things get tough, pay close attention to who shows up for you and who doesn't.
- **From conflict to connection**: Anger motivates us to stand up for our needs but can also cause biased thinking.
  - **Offense rumination**: Endlessly replaying negative events or harboring revenge fantasies.
  - **Bad guys bias**: Casting ourselves as the righteous hero and those who've wronged us as evil geniuses.
  - **Exploration network**: The brain region that helps us explore, imagine, and innovate.
    - **Exploratory dialogue**: What if I'm wrong?
    - **Creative perspective taking**: Brainstorm as many crazy, creative reasons the person did what they did as you can.
- **Connecting to something greater than ourselves**
  - **Spirituality**: (Re)discovering the sacred in our daily lives, however that looks to us.
  - **Awe**: Feeling like something is vast, which helps us transcend our own perspective and connect to something greater than ourselves.

Thwarted Connection: Sample Shadow Habits

Category	Shadow Goals	Sample Shadow Habits
Protect: Fighting back to avoid guilt, shame, or blame.	• **Spite**: A drive to hurt others who've hurt us. • **Aggression**: A drive to lash out.	• Offense rumination: Endlessly rehashing the trigger. • Bad guys bias: Deciding we're the heroes and they're the villains. • Criticizing or judging others' actions or behaviors. • Excluding, marginalizing, or hurting others to feel better. • Trying to avenge, sabotage, or get even with those who've wronged us. • Hostility, even to neutral parties.
Prove: Seeking external evidence about ourselves or our place in the world.	• **Popularity**: A drive to be liked and socially powerful. • **Validation**: A drive to be accepted by others.	• Flaunting status symbols. • Becoming focused on fame (e.g., likes, followers). • Engaging in gossip or spreading rumors to prove social currency. • Needing social recognition or praise for our contributions. • Embellishing our achievements. • Associating with higher-status people to elevate our social standing. • Using social connections for personal gain. • Frequently mentioning connections with influential or well-known people. • Trying to win important people over with flattery and compliments.
Prevent: Denying experiences, or devaluing our needs, to feel better right now.	• **Pretending**: A drive to avoid rocking the boat by wearing a social mask. • **Seclusion**: A drive to close ourselves off from others.	• Ignoring feelings of hurt or rejection. • Sacrificing our needs or boundaries to maintain relationships. • Justifying others' harmful behavior. • Shying away from confrontation by doing nothing or staying silent. • Conforming with others' opinions or expectations (including agreeing to unwanted requests). • Withdrawing or detaching to protect against further hurt. • Choosing solitary activities over social gatherings, events, or interactions. • Frequently deciding people are "dead to us."

Conclusion

Building a Shatterproof Life

> "It is better to light one small candle than to curse the darkness."
> —Eleanor Roosevelt

In the heart of Highland Park, a once-thriving Detroit suburb now ravaged by neglect, Shamayim Harris—a public school administrator, mother of four, minister, and reserve police officer known in her community as Mama Shu—would soon confront a heartbreak that would change her life forever.

Back in 1910, Highland Park became home to Ford Motor Company's first factory. The millions of Model T cars it produced birthed the middle class, creating an economic engine that built schools, hospitals, and homes, which became the envy of the world. But by the late twentieth century, the auto industry had fled, and the once-prosperous city fell into decline. Residents relocated, tax revenue tanked, and crime soared.

By the early 2000s, the situation in the predominantly Black community was dire. Mama Shu will never forget the time the power company repossessed all of Highland Park's streetlights, plunging the city into literal and figurative darkness. But tragically, even darker days were still to come.

On September 23, 2007, while Mama Shu and her husband were at work, their two boys—two-year-old Jakobi Ra and ten-year-old Chinyelu—were playing outside under a neighbor's supervision. Holding hands as they'd been taught, the brothers had just stepped off the curb to cross the street when a speeding car struck Jakobi, sending his tiny body hurtling

Conclusion

through the air—then it drove off. At the hospital, Jakobi was pronounced brain dead. The devastated family gathered at his bedside before removing him from life support.

From that moment, Mama Shu was trapped in an unfathomable nightmare. Each breath felt like swallowing shards of glass. Later that night, she feared her heart might simply stop beating. As days turned to weeks, the depth of their loss set in. Jakobi's toys gathered dust, and the vibrant spirit that had once filled their home was replaced with an eerie silence.

And yet, through her grief, Mama Shu somehow found fresh courage to chart a new course. Every day on her way to work, she kept noticing one desolate block. Beyond the boarded-up homes, weeds, and trash, she pictured how beautiful Avalon Street could be. That's when it hit her: maybe she could work to heal the broken parts of her suffering community and, in so doing, mend her own. She decided to begin by pouring every ounce of energy she had into transforming this single blighted block.

Six months after Jakobi's passing, a house on the corner of Avalon and Woodward went up for sale. Pooling her tax refund with a loan from her sister, she offered $3,000, and Avalon Village was born. In Jakobi's memory, Mama Shu would create a beacon of hope and healing for her community and a haven for its children, 60 percent of whom lived below the poverty line. She envisioned the city rising like a phoenix, again becoming a place where flowers bloomed, children played,[1] and people thrived.

Her mission soon attracted donations, which allowed her to slowly acquire more lots. With every nail hammered and weed she and her volunteers pulled, Mama Shu found strength in her community, and they found strength in her. To achieve her ultimate vision of a block that had everything the community needed, she founded the nonprofit Avalon Village. For their inaugural fundraiser, her team suggested a $50,000 goal, but Mama Shu set her sights higher. After their campaign went

---

1. If you caught the reference to Jonathan Larson's achingly beautiful *Rent*, I see you.

Conclusion

viral, Avalon Village met their $250,000 goal in just thirty days, and Mama Shu found herself feeling hopeful again. The first priority for the funds was completing the Homework House, a sanctuary that would give children a safe place where they could study, enjoy a nourishing meal, access shower and laundry facilities, and find solace in support groups for those who'd lost family members to violence.

Fourteen years later, the once-blighted block was unrecognizable. Mama Shu and her dedicated volunteers had succeeded in revitalizing Avalon Street—in the process, she had transformed her grief into "something bearable and something beautiful."

Then, on the evening of January 26, 2021, tragedy struck once more, in the cruelest way imaginable. Mama Shu's eldest son, twenty-three-year-old Chinyelu—the one who had been holding his brother's hand on that fateful day—was working security in Avalon Village. Patrolling the area from his truck, he'd parked in a grassy lot facing his mother's home. As snow softly fell, two assailants approached his vehicle and fired five shots into the car at point-blank range. Despite his wounds, he managed to crawl to a neighbor's doorway but help didn't get there in time.

That night, Mama Shu lost her only surviving son. The pain was like getting struck by lightning twice, familiar yet utterly indescribable. Grief-stricken and overwhelmed, she briefly considered giving up. But in the days that followed, her anguish only added to her resolve. Vowing that no other mother would experience such suffering on her watch, she and her team immediately began mobilizing a security team, erecting fences, installing cameras, and continuing their revitalization work.

A year to the day after Chinyelu's death, Mama Shu finally received the occupancy certificate for her Homework House. As she celebrated her proudest accomplishment five years in the making, she could feel Chinyelu's spirit rejoicing alongside her.

Today, Avalon Village spans three blocks with forty-five lots and six structures, including a Village Hall dedicated by Ellen DeGeneres. Once-littered yards have blossomed into playgrounds, tennis courts, and a flourishing community garden supplying organic produce to residents. Three solar-powered shipping containers house the Goddess Marketplace

Conclusion

(a women-owned business incubator), a STEAM lab,[2] and even a wine-tasting room. The Homework House boasts a library, computers, a 3D printer, and a music studio, while the newer Healing House offers yoga, Reiki, and massage. And Avalon Village was Highland Park's first area to be relit by five Wi-Fi-enabled solar streetlights.

At Jakobi Ra Park, the community gathers for concerts and picnics beneath a towering tree; at its base, white petunias frame a photo and plaque honoring Jakobi. Residents can find a few moments of peace and solace in the serene landscape of Chinyelu's Invincible Garden, or play a pickup game of basketball on a court with a brightly painted mural where Mama Shu's sons' faces stretch across the asphalt. And she still isn't finished. She wants to expand amenities and resources, construct market-rate housing, and beautify even more blocks "throughout the whole city."

While this external transformation is inspiring, Mama Shu's inner transformation has been equally profound. By harnessing her broken heart to heal her community and herself, she explains that she turned "grief into glory, pain into purpose, and loss into love." But lest anyone romanticize this journey, she's quick to point out that life has dealt her "some raw shit," and she'll forever carry the pain of losing her sons.

In this reality lies one essential truth: **becoming shatterproof doesn't mean never breaking—it means continually choosing to grow forward even in the face of devastating setbacks.** As Mama Shu explains, "You can live with a broken heart, but it doesn't have to break your spirit." The truth is, on our shatterproof journey, new storms will keep coming—and instead of hiding below deck, we must summon new resolve to adjust our sails. Indeed, as Mama Shu wisely notes, continuing the voyage ultimately "does end up being a choice."

BUILDING YOUR BEST SELF

As you embark on your shatterproof journey, I want to leave you with a few tips to navigate the choppy waters you might encounter. Let me

---

2. Which stands for science, technology, engineering, the arts, and math.

assure you that I didn't just dream up these insights one day at my laptop. I had to earn them, usually by making the wrong moves before I stumbled upon the right ones.

The first tip, that **it's okay to not be okay,** may sound familiar. But let's be honest: intellectually grasping it is one thing, but wholeheartedly internalizing[3] it is quite another. It's taken me years to get there. During that time, I, of course, had no trouble preaching this wisdom to others (clients, friends, unsuspecting strangers on airplanes), but I couldn't seem to apply it to myself. The breakthrough came when I finally saw the relentless drumbeat of grit gaslighting and toxic positivity for what they were—harmful pressure to put other people's comfort over our own well-being. And it was only when I embraced *not being okay* that I could find new ways to *actually be okay*. Embracing our pain is the first, and sometimes hardest, step in this journey, and the courage to take it will yield big rewards.

The next tip is even more foundational. **Prioritizing your needs isn't selfish or indulgent; it's an essential ingredient for a better life.** Especially in tough times, actively shaping your life around confidence, choice, and connection will bring out your most positive and powerful tendencies. Through enhanced energy, joy, and well-being, you'll start *feeling better*. Through stronger performance and deeper learning, you'll start *doing better*. And by charting a life based on your fundamental psychological needs, you'll soon be *living better*—with greater meaning,[4] purpose, peace, and even physical health and longevity.

In short, you'll consistently become the best version of yourself, which in turn, will benefit everyone around you. Indeed, pursuing your own needs isn't a zero-sum game. As I often remind my CEOs, "When *you* get better, everyone benefits." The same is true for the rest of us.

---

3. For my science fiction nerds, an even better word here is "grocking" (i.e., to understand something deeply and intuitively, with profound insight and empathy).

4. One interesting data point: Research has shown that the search for meaning is negatively related to the experience of meaning, suggesting that many common approaches to finding it aren't very effective (that's the beauty of need crafting: it actually works!) (Michael F. Steger et al., "Understanding the Search for Meaning in Life: Personality, Cognitive Style, and the Dynamic between Seeking and Experiencing Meaning," *Journal of Personality* 76, no. 2 [2008]: 199–228).

Prioritizing your needs will make you a better parent, partner, friend, worker, colleague, and citizen. You'll become less emotionally reactive and judgmental. You'll find yourself being more generous with others. It's a classic win-win.

The important thing is to identify what you want your life to look like and give yourself permission to go after it. Psychologists call this "**life crafting**." Broader than crafting your needs in any specific situation, it means taking a big step back to define what's important, and what's *most* important, to you. This exercise can be surprisingly simple. Check out the following list and circle (or add) the four to five most important elements of *your* life. Then, rank them from most to least important. This will help you determine what you can compromise on and what you can't:

- Physical health
- Mental health
- Work/job
- Career
- Marriage/romantic relationships
- Kids
- Extended family
- Friends
- Personal growth
- Creative pursuits
- Leisure/hobbies
- Spirituality/religion
- Volunteering
- Community

Finally, we should remember that **who we are in the future doesn't have to be who we were in the past.** After decades spent helping executives transform how they lead and live—followed by relearning all of these lessons in my own life—I can conclusively state that it *is* possible to change our reactions, our goals, our beliefs, and our habits. We can even

Conclusion

change our fundamental traits. My longtime editor told me not to mention any new research in this conclusion, but I'm going rogue because this is too important (sorry, dearest Talia). Here's the scoop: Studies show that our core traits aren't set in stone—they can change significantly over time, and for the better. In one study, personal growth didn't just predict improved happiness; it was *twice as effective* as any other factor they studied, including career success and marital status, *combined*.

In my last book, *Insight*, I argued that self-awareness is the metaskill of the twenty-first century. That's still true, and it's becoming increasingly clear that the metahabit of the twenty-first century must be transformation. So please, don't put off building your best possible self and your best possible life.

STAYING THE COURSE

Once your shatterproof journey starts gaining momentum, there are two common traps we can run into that lead us astray. First, beware the **overload trap.** Once you've probed your pain, traced your triggers, spotted your shadows, and picked your pivot, it's tempting—especially for stressed-out strivers—to take on too much. This usually involves choosing more than one shatterproof goal or making your list of shatterproof habits too long or complicated. But the outcome of setting unreachable standards is inevitable: You'll feel defeated, cynical, or give up altogether, and your shatterproof journey will be over before it really starts.

Remember my earlier advice that the secret to successful personal growth is to radically simplify it. When I coach CEOs, we pick one singular goal to work on *for one entire year*. Sometimes, we make enough progress that we eventually add a second, but we never begin with more than one at a time. The reason we do this? Because it works. (It works so well that I offer a guarantee: If a client doesn't change, I'll refund their money.) For my fellow overachievers, simplifying your development might feel like slacking, but it's actually the surest source of successful shatterproof transformations. Don't let perfection be the enemy of progress—it is through small victories that we win the game.

The second trap appears once we begin testing our shatterproof habits. During strategic experiments, we might feel like a fly trapped on the sticky traps of our shadow habits—mightily struggling to break free but unable to escape them. When we're caught in the **inertia trap,** we willingly surrender our power to what psychologist Stephen Hayes[5] calls "the dictator within." This inner tyrant confines us to our comfort zone, robbing us of our autonomy while claiming it's for our own good. Yet when we find ourselves avoiding an action simply because it feels unpleasant, this can actually be a sign that it's the action we *most* need to take.

During the time that my illness started worsening, I did something that "past Tasha" never would have—I let my inner dictator convince me that even light exercise would make things worse. Each day I avoided it, my aversion grew. Eventually, I stopped moving entirely before becoming bedbound. Then when Dr. Schofield diagnosed me, she shared the news I least wanted to hear: one of the best ways to manage EDS pain and stave off disease progression was daily, gentle exercise.[6] So, I just had to do the exact opposite thing my mind was telling me to do. *No problem.*

Fortunately, Stephen Hayes offers a great tool for these moments. The **reverse compass** helps us unravel our deepest-seated shadow habits by "pok[ing] the dictator within." If you feel overpowered, simply identify a value, goal, or principle that your dictator would hate. In my case, it was leaving the comfort of my zero-G bed and getting back in shape. Every day at 4:00 p.m.—my designated workout time—I had a chat with myself. *If you keep opting out, your pain will get worse, and soon, you'll never be able to get out of bed.* (I later shortened my pep talk to the melodramatic yet motivating mantra, "Stop moving, start dying.")

Focusing on the price I'd pay, rather than how tired I was and how scary exercise seemed, helped me overcome my literal inertia and reconnect

---

5. Dr. Hayes is one of the most well-respected pioneers of acceptance and commitment therapy (ACT), which we learned about in Chapter 5.

6. I want to be careful here. So many people with chronic or undiagnosed illnesses are bombarded with advice from well-meaning people that their problems will be solved through yoga, a vegan diet, or exercise. Exercise, for one, can be dangerous for some diseases (like ME/CFS), which just underscores how important fighting for an accurate diagnosis really is.

Conclusion

my behavior with my shatterproof goal of health.[7] That's exactly the point of the reverse compass: to defy our inner dictator and align our actions with who we truly want to be.

WHAT NOW?

We've certainly covered a lot of ground—now, where should you go from here? What should you do as soon as you put down this book? Lest I leave you hanging, in addition to the resources and tools we've already covered, let me share three more.

The first tool is a web-based self-assessment to find out how close you are to your resilience ceiling and what that means for you. You can take the free Resilience Ceiling Quiz at www.Resilience-Quiz.com. And because others' views of us can be equally important as our self-views (and are often more accurate), there's an option to send the quiz to someone who knows you well to compare answers. Your report will offer practical suggestions to get started on your shatterproof journey.

What's more, you can check out two downloadable bonus chapters to help you apply the shatterproof concepts and roadmap at work: building shatterproof teams (for current and aspiring leaders) and building shatterproof organizations (for executives). Those bonus chapters can be found at www.shatterproof-book/bonus.

Finally, if you are someone who wants to deeply and comprehensively apply the concepts of this book to your own life, I invite you to check out the comprehensive Shatterproof Workbook, which you can purchase at www.shatterproof-book/workbook.

STEPPING INTO THE LIGHT

As Mama Shu's story underscores, our journey to become shatterproof will rarely be a straight line; it's more likely to resemble a spiral. Archetypes of

---

7. If this still doesn't work, I suggest Mel Robbins's five-second rule, where you decide you're going to get up and do the thing in the next five seconds, then count down: "Five . . . four . . . three . . . two . . . one. DO IT."

Conclusion

growth and transformation, spirals have appeared in nature for millennia—consider giant galaxies, the smallest seashells, or the paths that falcons fly.[8] As when we draw a spiral, we start our journey from a fixed point. Then, with each self-limiting shadow discovered and shatterproof habit forged, we move outward in a gradually expanding circle. Brief pauses may be needed from time to time, but when we keep traveling forward, we'll continue to widen and grow. We may still encounter familiar obstacles, but with each revolution, we return in an expanded form with new insight, confidence, and power.

Becoming shatterproof transforms your life from the inside out—turning your biggest fears into your greatest opportunities, your toughest times into your proudest moments, and your shrinking reserves of resilience into a renewable source of strength. Even in moments where you've lost something precious you can't get back, or a part of your life is damaged beyond repair, you'll be able to compassionately accept your pain, conjure the determination to grow in spite of your circumstances, and build a new, good life.

When we rely on resilience to get us through our toughest challenges, we spend so much energy and effort just trying not to break. But when we become shatterproof, we realize that **it's okay to bend, to break, or even to hit rock bottom, knowing that with the right tools, in the times that break us, we can uniquely remake us.** Of course, setbacks will test your confidence and resolve. Embrace them. Learn from them. Let them crack or break you. Then pick up the pieces and repair them with precious metal. Before you know it, you'll be better and stronger than ever.

We began Chapter 1 with a quote from my favorite TV show, *Buffy the Vampire Slayer*: "The hardest thing in this world is to live in it." Let's briefly return to that scene. Buffy and her younger sister, Dawn, are standing atop a tower dozens of stories in the air, facing a portal of swirling purple energy that's just opened into another world. The air crackles and the sky is a tumultuous canvas of dark clouds. With a mix of sorrow

---

8. Fun fact: The *Mona Lisa*, arguably the most famous painting on Earth, is a perfect Fibonacci spiral (The Fibonacci Sequence, "The Mona Lisa," accessed September 11, 2024, https://thefibonaccisequence.weebly.com/mona-lisa.html).

and determination, Buffy confides to her sister, "The hardest thing in this world is to live in it." At that moment, she realizes that true bravery isn't about fighting monsters or facing death but about making our way in this messy, beautiful, scary, and wondrous life.

Before she jumps into the portal to save the world, Buffy urges her sister, "Be brave—*live*."

Carl Jung once said, "In the soul there lives a longing for light and an indispensable urge to emerge from the . . . darkness." Finding the courage to emerge from the darkness is exactly what living a shatterproof life is all about. Because without heartbreak, we cannot feel joy. Without failure, we can't know success. And without tragedy, we cannot find transcendence.

As both a scientist and human being, here's what I've come to believe: far from being an inconvenient co-traveler, darkness can be our greatest motivator and our greatest teacher. This doesn't mean that we should welcome, idealize, or "pain shop" the adversity we face. It simply means that even when we're surrounded by darkness, we can still find the courage to grow forward, into the light.

And so, as you set off on your own shatterproof journey, consider this question: *What if you could become the best version of yourself, not in spite of your broken parts but because of them? What if you could be you, but better?*

Epilogue

It's been almost three years since Dr. Schofield handed me a gift, wrapped in medical jargon: the primary diagnosis of hypermobile Ehlers-Danlos syndrome, along with its cotravelers. Since then, my list of diagnoses has grown, with additions like common variable immunodeficiency, vascular insufficiency, and craniocervical instability. (I tell you, the fun never stops!)

This avalanche of diagnoses would have paralyzed "old Tasha." But now, they energize me. While EDS has no cure, I now understand which aspects of my health I can maximize. I've conducted hundreds of strategic experiments—some big, like a life-changing neurovascular surgery to open a 95 percent–blocked jugular vein; some moderate, like giving up gluten to reduce my inflammation; and many small, like trying new medications to manage my symptoms. Here, my strategy is simple: if something has a reasonable chance of making me feel even 1 percent better, I'm willing to try it. And while I can't know what tomorrow will bring, I currently feel better than I have in over a decade.

When I share this development, people often exclaim, "It's a miracle!" But you know what? It wasn't a miracle. *It was me.* By facing my reality, I was able to take proactive steps to feel as good as possible in what I jokingly call my "defective meatsuit."

Let me paint a picture of my new life as an EDS warrior. I've assembled a team of fifteen incredible medical professionals who help me, hear me, and appreciate my color-coded binders. Every day, I take fifty-two

Epilogue

pills to manage my symptoms, and every night, I get ten hours of sleep—because after years of telling myself that I could subsist on six, I've made peace with the fact that my body needs ten. Through gentle exercise and physical therapy, my daily pain has plummeted from a nine out of ten to a manageable (for me) six. Every weekend, I spend a few hours sticking three small needles into my legs to infuse the plasma of ten thousand people into my blood so my body can have more energy and fight infections (thank you, donors!). With the help of my bionic ears (aka hearing aids), I can now actually hear people and no longer have to nod along and pretend I know what they're saying. And after losing more than half my hair, I get to wear a gorgeous wig, giving me my best hair day, every day.

But this transformation hasn't just been physical. I have felt it ripple across every part of my life. Instead of tolerating poor treatment from loved ones, I've fired my barnacles and leaned into my backers who offer me grace, kindness, and support. I've learned to protect my energy by saying no without guilt or apology, maintaining my boundaries, and taking "recovery days" when needed. I remain committed to my own needs while appreciating the needs of others. The biggest difference? In all parts of my life, I am now consistently a person who—like the COO from Chapter 8 proclaimed—makes the choice to be an active participant in my own life.

As a result, I feel more like myself than I have in years *and* like a better version of myself. In my coaching, I've never been more connected to my clients or seen more dramatic results. In my speaking and writing, I've stopped hiding and started sharing my story; it has lifted an immense boulder off my shoulders that I didn't even realize I was carrying, and I love hearing feedback about how much it is helping others.

I've also channeled my newfound insight and energy into deeper need fulfillment in two big ways. First, twenty-three years after I left my first love in life—theater—to pursue a more "practical" path as an organizational psychologist, I'm now throwing myself into a second, simultaneous career as a Broadway investor and producer. I've realized that *my* greatest awe is found under the dimmed lights of a transcendent

Epilogue

Broadway musical—especially one that I helped bring to life.[1] I am certain I wouldn't have rediscovered this passion, or had the courage to rearrange my life to honor it, without the lessons I learned from my shatterproof journey.

Now for another big change. For many years, I've been part of Marshall Goldsmith's 100 Coaches Program, where my mentor (and the world's number one leadership thinker) decided to teach one hundred people everything he knows for free. His goal was to help us make a bigger positive difference, and the only condition was that we'd each start a group of our own someday. For years, I dwelled on the "someday" part. But when I started digging out from my illness, I kept coming back to one question: If not now, when? So I became the first of Marshall's 100 Coaches to start my own pay-it-forward program, the Tasha Ten. Today, we have thirteen active and six emeritus members representing eleven countries, all activating human-centered leadership around the world by building better businesses, institutions, and communities.[2] While my initial goal was simply to give back, I've found that I'm getting more in return than I could have imagined—we are a chosen family, and these global visionaries and wonderful humans inspire and uplift me every day.

So, what's been the overall impact on my life of this journey to become shatterproof? I can sum it up in five words: **new peace and new joy.** My health problems certainly haven't disappeared, and I have faced a few other challenges—with one related to my marriage currently bringing me to my knees in a way I could have never imagined. But at least 90 percent of the time, peaceful sleep has replaced restless nights; action has replaced crippling panic; and grace has replaced self-loathing (as for the other 10 percent—hey, I'm human, and you are too).

Wondering whether these victories were largely internal—that is, felt by me but less noticeable to others—I asked a few close friends if they'd

---

1. Today, I bought a dress for the Tony Awards after the first show I joined was nominated for Best Revival of a Musical! WHAT?!

2. If you'd like to learn more about the Tasha Ten—or even apply—go to www.tashaeurich.com/tashaten for more information.

Epilogue

seen any changes. At first, the difference wasn't always obvious. My friend Dana noted that the new me was "much less available to do things together" (thanks to the deadline for this book!). But digging deeper, she noted, "Dare I say, you are more self-aware. Plus, you're letting yourself feel your feelings and giving yourself more grace." She mentioned the night I mixed up the date of her daughter's eighth-grade continuation ceremony. Upon realizing my mistake, I immediately got in the car, wigless and without a stitch of makeup. "You showed up, imperfectly and wonderfully, in a way that mattered to others, who were none the wiser," she told me. My friend Emily observed, "You are more open to taking chances and appreciating the vibrancy of life and the things you love. And it has helped you find better strategies for tackling challenges."

Living with my chronic illness and disability, along with all of life's other surprises, isn't easy, but my shatterproof journey has given me new sources of purpose, meaning, and joy. In medical school, doctors are taught to expect to find a horse rather than a zebra when they hear hoofbeats—that is, for a patient with unexplainable symptoms, they should look for the most common answer rather than the most surprising one. People with rare illnesses spend decades searching for answers while being told we look "normal." But sometimes, when you hear hoofbeats, it really is a zebra—the term the EDS community has reclaimed to describe ourselves. And do you know what zoologists call a group of zebras? A *dazzle*.

A few months ago, I spoke with the mother of a teenage girl suspected of having EDS. Through heart-wrenching tears, she described her daughter's relentless pain, the doctors who dismissed her, and misguided treatments that were only worsening her condition. I could feel her agony through the phone. Then, I remembered the advice that had once lit a shatterproof fire in me. I told her that she and her daughter could still live a full, vibrant life. It was time to fight like they never had before—and emerge stronger for it.

The same is true for all of us. Whatever challenges we face, the solution is not to suffer in silence, barely holding on and keeping it together—it's to stare our pain in the face and fight, fight, *fight* for the dazzling life that lies ahead of us.

Acknowledgments

Writing *Shatterproof* has been a five-year whirlwind of manic excitement, unhinged meltdowns, and overwhelming gratitude for the incredible people who supported, challenged, and inspired me along the way.

First, to two extraordinary and powerful women: my editor, Talia Krohn, and my agent, Christy Fletcher. This book—and probably my writing career—wouldn't exist without your brilliance, steadfast support, and unshakable belief in me. I truly don't know how I got so lucky.

To the phenomenal teams at Little, Brown Spark (Lauren Ortiz, Katherine Akey, Pat Jalbert-Levine, Sofia Sanchez, Gregg Kulick, Lauren Hesse, Laura Essex, and Allison Gudenau) and UTA (Veronica Goldstein, Melissa Chinchillo, Gráinne Fox, Yona Levin, Claire Yoo, and Samori Cullum)—thank you for your passion and partnership in bringing this book to life. My gratitude also crosses the pond to Sheila Crowley at Curtis Brown and Ríbh Brownlee at Pan MacMillan.

To my SpeakersOffice family: Holli Catchpole, Michele Wallace, Tracey Bloom, Cassie Glasgow, Sandy Conner, Kim Stark, Jessica Case, and Jennifer Canzoneri, thank you for twelve years of *woo-hoo*s and a wildly wonderful ride. And a shoutout to Mike Ganino for helping me craft a keynote that brings this book's message to life.

To everyone who participated in our studies and shared your stories, you are truly the heart of this book; thank you for your courage and trust. Endless gratitude also goes to our research team (Janae McKendry, Roger Burleigh, Brittainy Charette, Abby McNamara, Elle McNamara, Andrea Palmer, Nicole Podpirka, Maggie Reichenbach, and Lauren Vanderburg) for standing up a massive research program in the middle of a global pandemic—and crushing it.

Acknowledgments

To the brilliant minds who shaped *Shatterproof*: Thanks to Gareth Cook, Kate Rodman, and Will Storr for your invaluable expertise and guidance. To Michael Palgon, who patiently weathered every version of this book and reassured me that the fourth time *really* was the charm. To Adam Grant, your feedback on this book's framing was invaluable. To my friends and mentors for their generous guidance: Marshall Goldsmith, Hortense le Gentil, Laura Gassner Otting, Liz Wiseman, Michael Bungay Stanier, Stephanie Szostak, Tiffany Dufu, Tiffani Bova, Chip Heath, Alan Mulally, Amy Cuddy, Sarah MacArthur, Pam Sherman, Srdja Popovic, Scott Gurka, Erik Spoelstra, Caroline Webb, Chester Elton, Mark Thompson, Bonita Thompson, Hubert Joly, Alex Osterwalder, Chris Porath, David Burkus, and David Ladek.

To our early feedback givers and brilliant beta readers—Barry Engelhardt, Amy Riley, Lisa LaBrecque, Rhonda St. Croix, Edwin Palsma, Omar Ahmed, Bruce Morgan, Tricia Fitzpatrick, Jennifer Sanford, Udo Lange, Sue Gilad, Ashley Wieland, Mike Bearup, Barbara Wankoff, Gerson Elias, Emily Scott, and Allison Monkman—your insights made this book sharper, richer, and better in ways I never could have without you.

To my incredible team at The Eurich Group (Jasmine Lawrence, John Pugh, Danielle La Rose, Cynthia Shetty, Lisa Donmall-Reeve, and Michelle Longmire), thank you for keeping everything running so I can dive headfirst into wild projects like this—and for your patience with my comically slow email responses this year!

To my (chosen) family: Terri Wanger, Richard Wanger, Sarah Gibson Daly, Dana Sednek, Berkeley Bowler, Nick Baez, Roger Burleigh, Marshall Goldsmith, Hortense le Gentil, Hubert Joly, Laura Gassner Otting, Emily Wasserman, Bryan Daly, Darren Jeppson, Jerraud Coleman, Coles Whalen, Dave Gaddas, Rachel Maxwell, and Austin Leighty (. . . always)—and to the Tasha Ten. Thank you for grounding me, cheering me on, and forgiving me (I think) for disappearing into book jail for an eternity. Your love and belief have carried me further than I ever could have imagined.

And finally, to those of you who picked up this book—thank you for giving *Shatterproof* a place in your life. A wise friend once told me, "What

Acknowledgments

I wish for you for this book is that you can finally be completely and truly yourself. If you can, you will help your readers more than you can possibly imagine." I truly hope *Shatterproof* lives up to that wish—giving you strength to embrace your cracks, courage to navigate the chaos, and light even in your darkest moments.

Appendix A:
Crisis Mental Health Resources

Focus	Resource	Contact	Languages	Deaf and Hard of Hearing
Suicide, General Crisis Support	Suicide and Crisis Lifeline (US & Canada)	• Call or text 988 • More resources: https://988lifeline.org/	• English • Spanish (US) • French (Canada)	• TTY Users: Use preferred relay service or dial 711 then 988 • ASL Videophone Info: https://988lifeline.org/deaf-hard-of-hearing-hearing loss/
	Crisis TextLine (US)	• Text "HOME" to 741741 • Chat, WhatsApp, more resources: https://www.crisistextline.org/	• English • Spanish	• N/A
	Crisis Services (Canada)	• Call 1-833-456-4566 • Text 45645 (4:00 p.m.–midnight ET)	• English • French	• TTY Users: Use your preferred relay service or dial 711
Substance Abuse	Substance Abuse and Mental Health Services Hotline, also for family members (US)	• Call 1-800-662-HELP (4357) • Text 435748 (HELP4U) • More resources: https://www.samhsa.gov/find-help/national-helpline	• English • Spanish (phone only)	• TTY Users: 1-800-487-4889
	Canadian Centre on Substance Abuse (Canada)	• For a list of treatment hotlines, visit: https://www.ccsa.ca/addictions-treatment-helplines-canada • More resources: https://www.ccsa.ca/	• Varies	• Varies

Domestic Violence or Assault	National Domestic Violence Hotline (US)	• Call: 1-800-799-SAFE (7233) • Text: "START" to 88788 • Chat and more resources: https://www.thehotline.org	• English • Spanish • Access to interpreters in 140 languages	• Video phone: 1-855-812-1001 • Instant messenger: DeafHotline
	Domestic Violence Hotline (Canada)	• Call: 1-800-363-9010 • Chat: https://www.sheltersafe.ca/chat/	• English • French	• Use chat option
	National Sexual Assault Hotline (US)	• Call (800) 656-4673 • Chat: https://hotline.rainn.org/online • More resources: https://www.rainn.org/resources	• Spanish online chat: https://hotline.rainn.org/es	• Use chat option
	Sexual Assault Hotline (Canada)	• Call 1-866-644-3574 • Chat: https://www.awhl.org/	• English • French	• Use chat option

Appendix B:
What Are Your Resilience Risks?

Below are several experiences that could reduce your coping capacity through no fault of your own. The intention of this assessment is to help you have more self-compassion toward the adversity you've experienced and discover how it might impact your resilience. It's not a tool to predict whether you will be resilient or not but rather to help you better understand the risks that can make resilience harder.

For each list, place a checkmark next to the situations that apply to you.

LIST A: DAILY STRESS

Currently, I am experiencing . . .
1. Chronic stress, even minor stress
2. Health concerns, even minor ones (colds, flu, tiredness)
3. Ongoing job or career stress
4. A job change or loss
5. Discrimination
6. Mental health challenges
7. Problems with friends, colleagues, or family
8. Marital or romantic challenges
9. Loved ones experiencing significant problems

Appendix B: What Are Your Resilience Risks?

LIST B: LIFE STRESS

At any point over the last several years, I have experienced . . .
1. Financial problems or chronic financial stress
2. Legal trouble
3. A serious conflict with a close friend
4. Marital separation or divorce
5. A significant illness, injury, or accident
6. A disability or chronic illness
7. Being the victim of a crime
8. Abuse or assault (or serious threats of either)
9. Witnessing assault or violence
10. The death or illness of a loved one

LIST C: CHILDHOOD STRESS

As a child, I experienced . . .
1. Parental coldness or neglect
2. Parental divorce or death
3. Parental mental illness or substance abuse
4. Bullying or mistreatment from others
5. Economic hardship
6. Serious health issues or illness
7. Physical or sexual abuse
8. Witnessing violence or victimization of others

Sum up your score:
- For Lists A and B, give yourself 1 point for each selected item.
- For List C, give yourself 2 points for each item.
- Your final score will be a number between 0 and 36.

Appendix B: What Are Your Resilience Risks?

Total Score	What It Means
0–12	**Lower Lifetime Risk**: This score suggests you've faced relatively fewer risk factors affecting your resilience. While this likely means you have a longer "resilience runway," it doesn't mean you're immune to stress and adversity. Continue to nurture your mental resources and pay attention to new challenges that may be depleting your resilience.
13–24	**Moderate Lifetime Risk**: Your score means you've encountered a moderate number of risk factors that may be limiting your resilience, at least to some extent. Have self-compassion for how these experiences have shaped you, remember that it's okay to ask for help, and develop alternative coping strategies that don't exclusively rely on resilience.
25–36	**Higher Lifetime Risk**: Your score suggests a high number of risk factors in your life. Having faced substantial adversity, you might find it difficult in some situations to summon the resilience you'd like. Keep remembering two things: first, that these experiences are not your fault; and second, that acknowledging these risks and stressors is actually the first path to transformative growth.

Appendix C: Need Audit

	First, find your FIXATIONS	Second, uncover your FEARS
Confidence	Situational cues that . . . • We are ineffective. • We are impostors. • Others think we're incompetent. • Others are superior to us.	Worthlessness • Am I useless? • Am I no good? • Am I letting people down?
Choice	Situational cues that . . . • We're helpless. • We're being micromanaged. • We're doing things against our will. • We have to hide our true selves.	Powerlessness • Am I helpless over my life? • Do I even know myself at all? • Will I ever be happy?
Connection	Situational cues that . . . • We're left out or excluded. • People are angry or upset with us. • We aren't receiving the support we need. • Others are letting us down, being cruel, or betraying us.	Being unloved/unlovable • Does anyone really see me? • Does anyone really care about me? • Am I worthy of love?

Appendix D: How Shatterproof Are You?

For each of the following questions, use the rating scale below to select the option that best describes your current beliefs, actions, and habits for navigating challenges in your life.

1	2	3	4	5
Never	Rarely	Often	Usually	Almost always

RESILIENCE LIMITS

1. Through adversity, I prioritize growing forward instead of "just" bouncing back.
2. When my best coping tools aren't working, I change course.
3. I can spot the signs that my resilience reserves are running low.
4. I use my resilience strategically rather than as a one-size-fits all solution.

PROBE PAIN

5. I don't let a fear of breaking stop me from examining my negative emotions/pain.
6. I can identify toxic positivity from others.
7. I am able to identify when my freeze-or-faint system has taken over.
8. I have one or more tools I regularly use that help me probe my pain.

TRACE TRIGGERS

9. In any given situation, I can name my triggers.
10. When I feel triggered, I can identify the unmet need behind my reaction.
11. I understand my past triggers and how they influence my emotions in the present.
12. I conduct a weekly three-minute mood map to understand my emotional trends.

SPOT SHADOWS

13. I do whatever I can to prioritize intrinsic motivation over extrinsic motivation in my life.
14. I am able to notice when I'm pursuing quick fixes for self-worth, power, or approval.
15. When I'm acting out of character, I ask, "How is this behavior different from when I'm at my best?"
16. I am able to examine my shadows without getting stuck in self-judgment or shame.

PICK PIVOTS

17. When the status quo is no longer serving me, I take proactive action to change.
18. When my needs are being frustrated, I find different ways to get them met.
19. When I'm pivoting, I work on one clear shatterproof goal.
20. I use strategic experiments to find new ways to increase my need fulfillment.

Appendix D: How Shatterproof Are You?

Total Score	What It Means
20–45	**Less Developed Shatterproof Practices**: Your score suggests that you have an opportunity to make shatterproof practices a way of life. To get started, one idea is to focus on the three mind shifts in Chapter 4. You can also pay attention to the steps of the Shatterproof Road Map where your scores were lower. Before you know it, you'll be on your way!
46–75	**Developing Shatterproof Practices**: Your score suggests that you are well on your way to becoming shatterproof—and that you have the opportunity to strengthen certain steps in the roadmap or more consistently practice all steps in the process. One suggestion is to pick one step to prioritize in the coming weeks, then flip back to the tips and tricks we've covered for that step.
76–100	**Strong Shatterproof Practices**: Your score reflects a strong practice of being shatterproof. You frequently recognize the signals your pain is sending, understand your triggers and unmet needs, identify your shadows, choose what to change, and don't overly rely on resilience. Continue to build on these strengths, and you'll keep growing forward through hard times.

Appendix E: What Are Your Perfectionist Tendencies?

Please place a checkmark next to the thoughts, feelings, and behaviors you find yourself engaging in on a regular basis.

CONCERN OVER MISTAKES

1. If I fail at work/school, I feel like a failure as a person.
2. I hate being less than the best at things.
3. People will probably think less of me if I make a mistake.
4. If I do not do well all the time, people will not respect me.

DOUBT ABOUT ACTIONS

5. Even when I do something very carefully, I often feel that it's not quite right.
6. I usually have doubts about the simple everyday things I do.
7. I tend to get behind in my work because I repeat things over and over.
8. It takes me a long time to do something "right."

PERSONAL STANDARDS

9. If I don't set the highest standards, I am likely to end up a second-rate person.
10. I set higher goals than most people.

Appendix E: What Are Your Perfectionist Tendencies?

11. I am very good at focusing my efforts on attaining a goal.
12. Other people seem to accept lower standards than I do.

ORGANIZATION

13. I am an organized person.
14. In my life, everything has its place.
15. Neatness is very important to me.

To tally your score, give yourself 1 point for each checkmark. Your final score will be a number between 0 and 15.

Total Score	What It Means
0–4	**Low Perfectionist Tendencies**: You tend to set realistic goals for yourself and are comfortable with your imperfections. Because you don't need everything to be perfect, you can stay more self-accepting and mentally healthy.
5–10	**Moderate Perfectionist Tendencies**: You may set high standards and strive for excellence, but you don't hold yourself to these standards in all situations. It will be crucial for you to identify when perfectionism becomes counterproductive. Balancing high aspirations with self-acceptance will help you achieve your goals while protecting your well-being.
11–15	**Strong Perfectionist Tendencies**: Your score suggests strong perfectionist tendencies that could be significantly impacting your mental health and well-being. While striving for excellence can be helpful, excessively high standards lead to anxiety, self-criticism, and self-doubt. Pick a few of the tools from this chapter to start experimenting with to help you develop healthier goals and habits for personal success.

This scale is a sampling of items (some paraphrased) from four scales of the Frost Multidimensional Perfectionism Scale (excluding scales for parental expectations and parental criticism—because let's not open *that* Pandora's box today!).

Appendix F:
Paint-by-Numbers Reflected Best Self

STEP 1: PICK YOUR PEOPLE (<10 MINUTES)[1]

Identify ten to twenty people who know you well, from different parts of your life: colleagues (current/former), customers, friends (old/new), family members, anyone else you regularly interact with. (The more diverse this group is, the better your data will be.)

STEP 2: SEND THE FEEDBACK REQUEST (<10 MINUTES)

Send an email asking them to each share examples of times they've seen you at your best. Here's a template:

> As part of my development, I am contacting X people who know me well to better understand when I'm at my best. Would you consider participating?
>
> With this exercise, all you need to do is think about our interactions and identify two to four examples of a time when I was at my best in your eyes, and write a few sentences using these prompts:
>
> - One of the ways that you add value and make important contributions is . . . *[In a sentence or a few words, say what you see me do.]*

---

1. Adapted from Robert E. Quinn, Jane E. Dutton, and Gretchen M. Spreitzer, "Reflected Best Self Exercise: Assignment and Instructions to Participants," Center for Positive Organizational Scholarship, Ross School of Business, University of Michigan, product B 1 (2003).

Appendix F: Paint-by-Numbers Reflected Best Self

- I think of the time that . . . *[Write one to two sentences about a specific memory of when you saw that strength in me: describe the situation, my behavior, and the impact it had on you/others.]*

Thank you for considering this request! If you are willing to help, I would appreciate your responses by [*deadline*]. I will be sure to tell you what I learned.

It's normal to feel awkward when you think about writing this email: remember that most people find this a transformational exercise, and your network will usually be thrilled to assist you.

STEP 3: IDENTIFY THEMES (<20 MINUTES)

- Read through your feedback.
- Start identifying the commonalities you're seeing using this template.
  - Theme: For example, creativity, empathy, influence.
  - List the specific behavioral examples that fall under each theme.
  - Interpretation: What does this say about your strengths?

STEP 4: COMPOSE YOUR REFLECTED BEST SELF-PORTRAIT (<20 MINUTES)

- Create a portrait of your best self from what you learned (and feel free to share it with the kind folks who gave you feedback!). Here's a short sample:
  - I'm a compassionate listener who offers support and encouragement. I don't let criticism affect me, nor do I dwell on past mistakes. Instead, I focus on what is possible.
  - I build strong relationships and help others feel valued and respected. I collaborate to find common ground and achieve shared goals. I inspire trust and loyalty by being authentic and reliable.

Appendix F: Paint-by-Numbers Reflected Best Self

- I'm skilled at problem-solving and resolving conflict. I approach challenges calmly and rationally, searching for root causes and constructive solutions. I facilitate open dialogue to build consensus.
- I positively impact people's lives by helping them achieve their goals. I apply my empathy and kindness to create lasting change and even build better communities.

Appendix G: How Fulfilled or Frustrated Is Your Confidence?

Consider how you're thinking, feeling, and behaving right now. Place a checkmark next to the statements that you feel apply to you at this moment in time.

CONFIDENCE SATISFACTION

1. I feel effective and capable.
2. I am doing things that I'm good at.
3. I am confident in my actions.
4. I generally feel a sense of accomplishment in what I do.
5. I feel like I can overcome obstacles.
6. I am learning interesting new skills.
7. People tell me I'm good at what I do.
8. I feel prepared for future challenges.

CONFIDENCE FRUSTRATION

1. I experience situations where I am made to feel ineffective.
2. People tell me things that make me feel incompetent.
3. People have unrealistic expectations of me.
4. I experience situations where I feel inferior.
5. I don't have many opportunities to fulfill my potential.

Appendix G: How Fulfilled or Frustrated Is Your Confidence?

6. People imply that I am incompetent.
7. Other people doubt my capacity to improve.
8. I feel like a failure.

To tally your score, give yourself 1 point for each checkmark. You'll have two scores (Confidence Satisfaction and Confidence Frustration), each between 0 and 8. Map each score as follows:

- 0 to 2: Low fulfillment or frustration
- 3 to 5: Moderate fulfillment or frustration
- 6 to 8: High fulfillment or frustration

Finally, place an X in the box that reflects your scoring categories in the following matrix. The higher and more to the right you are, the better your well-being will generally be.

	Low Fulfillment	Moderate Fulfillment	High Fulfillment
Low Frustration			
Moderate Frustration			
High Frustration			

Appendix H:
How Fulfilled or Frustrated Is Your Choice?

Consider how you're thinking, feeling, and behaving right now. Place a checkmark next to the statements that you feel apply to you at this moment in time.

CHOICE SATISFACTION

1. I'm making decisions based on my authentic interests and values.
2. I am free to make my own choices.
3. I'm able to decide how to live my life.
4. I feel able to express my ideas and opinions.
5. I can pretty much be myself most days.
6. The choices I make express my true self.
7. I feel like I am free to choose for myself how to live my life.
8. I rarely feel totally helpless or powerless over my life.

CHOICE FRUSTRATION

1. I feel a lot of unwanted pressure or external demands.
2. I feel controlled by internal or external burdens.
3. I feel pressure to behave in certain ways.
4. I feel pushed to do things I don't want to do.
5. I feel prevented from making choices.
6. There are lots of people telling me what I have to do.

Appendix H: How Fulfilled or Frustrated Is Your Choice?

7. I'm forced to follow decisions or plans others make for me.
8. I do things because I have to, not because I want to.

To tally your score, give yourself 1 point for each checkmark. You'll have two scores between 0 and 8 (Choice Satisfaction and Choice Frustration). Categorize each score based on the following:

- 0 to 2: Low fulfillment or frustration
- 3 to 5: Moderate fulfillment or frustration
- 6 to 8: High fulfillment or frustration

Finally, place an X in the box that reflects your scoring categories in the following matrix. The higher and more to the right you are, the better your well-being will generally be.

	Low Fulfillment	**Moderate Fulfillment**	**High Fulfillment**
Low Frustration			
Moderate Frustration			
High Frustration			

Appendix I:
Are You Overidentified at Work?

Take a moment to reflect on the following questions. Answering honestly will help you understand whether you might be overidentified with your job.

1. How often do you think about your job outside of working hours?

1	2	3	4
Rarely	Sometimes	Frequently	Constantly

2. When meeting someone new, how quickly do you mention what you do for a living?

1	2	3	4
Very rarely	Only when asked	Early in conversations	Virtually immediately

3. With friends and family, how difficult is it to avoid talking about work?

1	2	3	4
Easy	Somewhat easy	Difficult	Extremely difficult

4. Do you have hobbies or interests outside of work that are not related to your professional skills?

1	2	3	4
I have more than two unrelated hobbies	I have at least one unrelated hobby	I've been meaning to find one	No, my work is my life

Appendix I: Are You Overidentified at Work?

5. How would you feel if you had to leave your job / career / profession tomorrow?

1	2	3	4
Accepting	Anxious	Quite upset	Absolutely inconsolable

Add up each number you chose—your final score will be between 5 and 20.

Total Score	What It Means
5–9	**Low Overidentification**: You have a balanced identity when it comes to work and the other aspects of your identity. You don't let your job become your life by engaging in diverse activities and disconnecting when needed/possible to experience more agency, less stress, and better well-being.
10–15	**Moderate Overidentification**: You identify with your work, but most likely not excessively. While you may think about work outside of office hours and discuss it often, you have other elements of your identity that are salient focus areas of your life. Don't forget to maintain this balance to avoid potential burnout.
16–20	**High Overidentification**: Your work significantly and disproportionately influences your identity, thoughts, and conversations outside of the workplace. While your company may appreciate this extraordinary dedication, it most likely isn't helping you (or even your employer) in the long run. This is a great time to work on building a more balanced identity: engaging in activities outside of work and diversifying your interests to supercharge need fulfillment and well-being.

Appendix J:
Offense Rumination and Bad Guys Bias: Your Tendencies

Think of the last time you experienced an "interpersonal injury"—that is, someone hurt or upset you. While thinking about your initial reaction to that situation, place a checkmark next to the responses you remember having.

1. I couldn't stop thinking about how this person wronged me.
2. I knew the other person was clearly out to get me.
3. My thoughts about this person's actions interfered with my enjoyment of life.
4. I decided I never knew the other person at all.
5. I had trouble getting thoughts about how I was treated out of my head.
6. I felt like the person was deliberately trying to ruin my life.
7. I replayed the event over and over in my mind.
8. I adamantly believed that they were an evil person.
9. I was preoccupied with thoughts of revenge.
10. I felt a sense of righteous indignation.

Tally the number of tick marks for odd- and even-numbered responses separately and use the following guide to discover your patterns around offense rumination and bad guys bias.

Appendix J: Offense Rumination and Bad Guys Bias: Your Tendencies

Total Number of Odd Responses =	Total Number of Even Responses =

Offense Rumination	**Bad Guys Bias**
0–1: In this situation, your offense rumination was low, which suggests you avoided unnecessary pain in the face of a disconnection trigger.	**0–1**: Your score suggests that you are less likely to rely on the biased belief that you are blameless and those who've hurt you are bad people.
2–3: In this situation, your offense rumination was moderate, which suggests that while you prevented some pain, you have more work to do.	**2–3**: Your score suggests that you may, at least somewhat, rely on the biased belief that you are blameless and those who've hurt you are bad people. With a bit more work, you should find yourself experiencing less offense rumination in no time.
4–5: In this situation, your offense rumination was high, suggesting you can greatly benefit from the tools in Chapter 11 to minimize preventable pain amid connection triggers.	**4–5**: Your score suggests a clear pattern of seeing yourself as blameless and those who have hurt you as bad people, which is likely feeding the fire of offense rumination and harming your well-being.

Appendix K: How Satisfied or Thwarted Is Your Connection?

Consider how you're thinking, feeling, and behaving right now. Place a checkmark next to the statements that you feel apply to you at this moment in time.

CONNECTION SATISFACTION

1. People are generally pretty friendly to me.
2. I get along with most people.
3. I consider the people I regularly interact with to be my friends.
4. People I interact with usually relate well to me.
5. I have a strong sense of intimacy and trust with those who matter to me.
6. I have a network that encourages me in stressful situations.
7. The people in my life give back to me what I give to them in care and concern.
8. I have at least one close and true friend I can confide in.

CONNECTION FRUSTRATION

1. The people I interact with regularly do not seem to like me much.
2. Other people can be dismissive of me or reject me.
3. I feel like I don't fit in.

Appendix K: How Satisfied or Thwarted Is Your Connection?

4. I often feel unsafe with or singled out by the people I interact with.
5. I pretty much keep to myself and don't have a lot of social contacts.
6. Important people in my life don't comfort me when I'm feeling low.
7. Important people in my life don't really care about me.
8. Sometimes the people I count on let me down.

To tally your score, give yourself 1 point for each checkmark. You'll have two scores between 0 and 8 (Connection Satisfaction and Connection Frustration). Categorize each score based on the following:

- 0 to 2: Low fulfillment or frustration
- 3 to 5: Moderate fulfillment or frustration
- 6 to 8: High fulfillment or frustration

Finally, place an X in the box that reflects your scoring categories in the following matrix. The higher and more to the right you are, the better your well-being will generally be.

	Low Fulfillment	Moderate Fulfillment	High Fulfillment
Low Frustration			
Moderate Frustration			
High Frustration			

Notes

Chapter 1: Welcome to the Chaos Era

15 **string of mild winters**: Crawford S. Holling, "Resilience and Stability of Ecological Systems," *Annual Review of Ecology and Systematics* 4, no. 1 (1973): 1–23.

15 **aggressive forest management and fire suppression**: David C. Powell, *Effects of the 1980s Western Spruce Budworm Outbreak on the Malheur National Forest in Northeastern Oregon*, USDA Forest Service, Pacific Northwest Region, Forest Insects and Diseases Group, 1994.

15 **over millions of acres**: C. A. Miller, "Spruce Budworm: How It Lives and What It Does," *Forestry Chronicle* 51, no. 4 (1975): 136–38.

15 **rained down on him like a shower**: Maine Forest Products Council, "Lessons Learned: Memories of the Maine Budworm Infestation, 1970s–1980s," posted on YouTube, April 1, 2015, https://www.youtube.com/watch?v=LRW9O-abhLY.

15 **twenty years' worth**: Boreal Forest Facts, "11 Big Questions about the Tiny Spruce Budworm," accessed September 9, 2024, https://www.borealforestfacts.com/?p=510.

16 **provincial governments and private landowners**: Maine SBW Task Force, "1970s–80s Outbreak," Center for Research on Sustainable Forests, University of Maine, accessed September 9, 2024, https://www.sprucebudwormmaine.org/historical-perspectivespast-infestations/1970s-80s-outbreak/.

16 **more than 3.5 million acres**: Maine Forest Products Council, "Lessons Learned."

16 **largest spruce budworm outbreak**: Maine SBW Task Force, "1970s–80s Outbreak."

16 **"phase transition events"**: John Thackara, "Limits of Resilience," Resilience, January 15, 2013, https://www.resilience.org/stories/2013-01-15/limits-of-resilience/.

16 **"compound extremes"**: Gabrielle Canon, "'Historic' Weather: Why a Cocktail of Natural Disasters Is Battering the US," *Guardian*, June 18, 2022, https://www.theguardian.com/us-news/2022/jun/17/compound-extremes-natural-disasters-us-west.

16 **chaos is accelerating**: Nicholas Bloom, Davide Furceri, and Hites Ahir, "Tracking Uncertainty in a Rapidly Changing Global Economic Outlook," VoxEU, December 17, 2022, Centre for Economic Policy Research, https://cepr.org/voxeu/columns/tracking-uncertainty-rapidly-changing-global-economic-outlook#:~:text=The%20measure%20suggests%20that%20uncertainty,associated%20cost%20of%20living%20crisi.

16 **uncertainty continues to rise**: The study by Bloom, Furceri, and Ahir found uncertainty increasing in more than seventy-one countries ("Tracking Uncertainty").

17 **working more but accomplishing less**: Jane Their, "American Worker Productivity Is Declining at the Fastest Rate in 75 Years—And It Could See CEOs Go to War against

WFH," *Fortune*, May 5, 2023, https://fortune.com/2023/05/05/remote-work-productivity-5-straight-quarters-decline-gregory-daco/.

17   **a staggering 75 percent**: Sophie Bethune, "Stress in America 2022: Concerned for the Future, Beset by Inflation," American Psychological Association, press release, October 2022, https://www.apa.org/news/press/releases/stress/2022/concerned-future-inflation.

17   **headaches**: Domenico D'amico et al., "Stress and Chronic Headache," *Journal of Headache and Pain* 1 (2000): S49–S52.

17   **inflammation**: Paul H. Black, "Stress and the Inflammatory Response: A Review of Neurogenic Inflammation," *Brain, Behavior, and Immunity* 16, no. 6 (2002): 622–53.

17   **immune problems**: Agnese Mariotti, "The Effects of Chronic Stress on Health: New Insights into the Molecular Mechanisms of Brain–Body Communication," *Future Science OA* 1, no. 3 (2015).

17   **cognitive issues**: John T. O'Brien, "The 'Glucocorticoid Cascade' Hypothesis in Man: Prolonged Stress May Cause Permanent Brain Damage," *British Journal of Psychiatry* 170, no. 3 (1997): 199–201.

17   **sleep disturbances**: P. Lavie, "Sleep Disturbances in the Wake of Traumatic Events," *New England Journal of Medicine* 345, no. 25 (2001): 1825–32; J. Philbert, P. Pichat, S. Beeske, M. Decobert, C. Belzung, and G. Griebel, "Acute Inescapable Stress Exposure Induces Long-Term Sleep Disturbances and Avoidance Behavior: A Mouse Model of Post-traumatic Stress Disorder (PTSD)," *Behavioural Brain Research* 221, no. 1 (2011): 149–54.

17   **depression**: Kristie T. Ota et al., "REDD1 Is Essential for Stress-Induced Synaptic Loss and Depressive Behavior," *Nature Medicine* 20, no. 5 (2014): 531–35.

17   **memory loss**: Guerry M. Peavy et al., "The Effects of Prolonged Stress and APOE Genotype on Memory and Cortisol in Older Adults," *Biological Psychiatry* 62, no. 5 (2007): 472–78.

17   **"emotionally coasting"**: Amy Beecham, "Emotional Coasting: Why Are So Many of Us Are [sic] Feeling Out of Sorts at the Moment?" Stylist, 2022, https://www.stylist.co.uk/health/mental-health/emotional-coasting/673895.

17   **"mustivation"**: Liesbet Boone et al., "Self-Critical Perfectionism and Binge Eating Symptoms: A Longitudinal Test of the Intervening Role of Psychological Need Frustration," *Journal of Counseling Psychology* 61, no. 3 (2014): 363.

17   **lonelier**: Tore Bonsaksen et al., "Associations between Social Media Use and Loneliness in a Cross-National Population: Do Motives for Social Media Use Matter?," *Health Psychology and Behavioral Medicine* 11, no. 1 (2023): 2158089.

17   **less time with our friends**: Simon Kaufman, "Why Are Americans Spending More Time Alone?," Scripps News, July 24, 2024, https://scrippsnews.com/stories/why-are-americans-spending-more-time-alone/.

18   **levels of life satisfaction**: Suzanne Blake, "Americans Are Becoming Less Satisfied with Their Lives," *Newsweek*, February 10, 2024, https://www.newsweek.com/americans-are-less-satisfied-their-lives-1868560.

18   **anxiety and depression**: Renee D. Goodwin et al., "Trends in Anxiety among Adults in the United States, 2008–2018: Rapid Increases among Young Adults," *Journal of Psychiatric Research* 130 (2020): 441–46; Inga D. Sigfusdottir et al., "Trends in Depressive Symptoms, Anxiety Symptoms and Visits to Healthcare Specialists: A National Study among Icelandic Adolescents," *Scandinavian Journal of Public Health* 36, no. 4 (2008): 361–36; Cindy Gordon, "Massive Health Wake Up Call: Depression and Anxiety Rates Have Increased by 25%

in the Past Year," *Forbes*, February 12, 2023, https://www.forbes.com/sites/cindygordon/2023/02/12/massive-health-wake-up-call-depression-and-anxiety-rates-have-increased-by-25-in-the-past-year/?sh=63c8e00d1760; Dan Witters, "U.S. Depression Rates Reach New Highs," Gallup, May 17, 2023, https://news.gallup.com/poll/505745/depression-rates-reach-new-highs.aspx#:~:text=Line%20chart%3A%20Rising%20trends%20in,most%20recent%20results%2C%20obtained%20Feb.

18 **fight-or-flight response system**: Leigh Cwart, "Stress Can Literally Kill You. Here's How," *Popular Science*, October 5, 2021, https://www.popsci.com/health/stress-effects-on-body/.

20 **kept our ancestors alive**: Roy F. Baumeister et al., "Bad Is Stronger Than Good," *Review of General Psychology* 5, no. 4 (2001): 323–70.

20 **four times more bad experiences**: Catrin Finkenauer and Bernard Rimé, "Keeping Emotional Memories Secret: Health and Subjective Well-Being When Emotions Are Not Shared," *Journal of Health Psychology* 3, no. 1 (1998): 47–58.

20 **pain of losing money**: Daniel Kahneman and Amos Tversky, "Choices, Values, and Frames," *American Psychologist* 39, no. 4 (1984): 341.

20 **dwell on negative information**: This is called the positive-negative asymmetry effect (Baumeister et al., "Bad Is Stronger Than Good").

20 **the pain of bad ones**: Kennon M. Sheldon, Richard Ryan, and Harry T. Reis, "What Makes for a Good Day? Competence and Autonomy in the Day and in the Person," *Personality and Social Psychology Bulletin* 22, no. 12 (1996): 1270–79.

20 **solitary traumatic event**: Baumeister et al., "Bad Is Stronger Than Good."

20 **stick to us like Velcro**: Lucy Hone, "The Three Secrets of Resilient People," TEDxChristchurch, August 2019, https://www.ted.com/talks/lucy_hone_the_three_secrets_of_resilient_people.

21 **drains the very resources**: Christy A. Denckla et al., "Psychological Resilience: An Update on Definitions, A Critical Appraisal, and Research Recommendations," *European Journal of Psychotraumatology* 11, no. 1 (2020): 1822064.

22 **Anne's stays awake**: Prolonged stress increases cortisol over time is shown in Shuhei Izawa et al., "Effects of Prolonged Stress on Salivary Cortisol and Dehydroepiandrosterone: A Study of a Two-Week Teaching Practice," *Psychoneuroendocrinology* 37, no. 6 (2012): 852–58. Even when chronically stressed people take dexamethasone, a cortisol-stopping hormone, their bodies don't actually stop producing it (Maren Wolfram et al., "Emotional Exhaustion and Overcommitment to Work Are Differentially Associated with Hypothalamus–Pituitary–Adrenal (HPA) Axis Responses to a Low-Dose ACTH1–24 (Synacthen) and Dexamethasone–CRH Test in Healthy School Teachers," *Stress* 16, no. 1 (2013): 54–64).

22 **Uncertainty involves unpredictable situations**: Dan W. Grupe and Jack B. Nitschke, "Uncertainty and Anticipation in Anxiety: An Integrated Neurobiological and Psychological Perspective," *Nature Reviews Neuroscience* 14, no. 7 (2013): 488–501.

22 **safer to assume the worst**: R. Nicholas Carleton, "Fear of the Unknown: One Fear to Rule Them All?," *Journal of Anxiety Disorders* 41 (2016): 5–21.

22 **uncertainty triggers our fight-or-flight system**: Julian F. Thayer et al., "A Meta-Analysis of Heart Rate Variability and Neuroimaging Studies: Implications for Heart Rate Variability as a Marker of Stress and Health," *Neuroscience & Biobehavioral Reviews* 36, no. 2 (2012): 747–56.

22 **massive toll on performance and well-being**: Bryan Robinson, "What Brain Science Reveals about Uncertainty and 6 Strategies to Cope at Work," Forbes, August 24, 2022,

https://www.forbes.com/sites/bryanrobinson/2022/08/24/what-brain-science-reveals-about-uncertainty-and-6-strategies-to-cope-at-work/?sh=5a4b90b244b0.

22 **threatens our sense of safety**: Carleton, "Fear of the Unknown."

22 **worry**: Markham Heid, "Science Explains Why Uncertainty Is So Hard on Our Brain," Medium, March 19, 2020, https://elemental.medium.com/science-explains-why-uncertainty-is-so-hard-on-our-brain-6ac75938662#:~:text=%E2%80%9CUncertainty%20acts%20like%20rocket%20fuel,in%20response%20to%20those%20threats.%E2%80%9D.

22 **anxiety**: Grupe and Nitschke, "Uncertainty and Anticipation in Anxiety."

22 **reactivity**: Christian Grillon et al., "Anxious Responses to Predictable and Unpredictable Aversive Events," *Behavioral Neuroscience* 118, no. 5 (2004): 916.

22 **desperate for answers**: Arie W. Kruglanski and Donna M. Webster, "Motivated Closing of the Mind: 'Seizing' and 'Freezing,'" *Psychological Review* 103, no. 2 (1996): 263.

22 **To cope, some people will claim certainty**: Irmak Olcaysoy Okten, Anton Gollwitzer, and Gabriele Oettingen, "When Knowledge Is Blinding: The Dangers of Being Certain about the Future during Uncertain Societal Events," *Personality and Individual Differences* 195 (2022): 111606.

22 **worrying about job loss**: Sarah A. Burgard, Jennie E. Brand, and James S. House, "Perceived Job Insecurity and Worker Health in the United States," *Social Science & Medicine* 69, no. 5 (2009): 777–85.

22 **"I don't even care what happens"**: Carleton, "Fear of the Unknown"; Wolfram Schultz et al., "Explicit Neural Signals Reflecting Reward Uncertainty," *Philosophical Transactions of the Royal Society B: Biological Sciences* 363, no. 1511 (2008): 3801–11.

22 **a study by neuroscientist Archy de Berker and colleagues**: De Berker et al., "Computations of Uncertainty," 10996.

22 **the fundamental human fear**: Carleton, "Fear of the Unknown."

CHAPTER 2: THE THREE MYTHS OF RESILIENCE

28 **social support**: Social support is often named the single most important protective factor for resilient outcomes. For a review, see Peggy A. Thoits, "Mechanisms Linking Social Ties and Support to Physical and Mental Health," *Journal of Health and Social Behavior* 52, no. 2 (2011): 145–61; Alexandra Stainton et al., "Resilience as a Multimodal Dynamic Process," *Early Intervention in Psychiatry* 13, no. 4 (2019): 725–32.

28 **positive emotions**: Ji Hee Lee et al., "Resilience: A Meta-Analytic Approach," *Journal of Counseling & Development* 91, no. 3 (2013): 269–79.

28 **grit**: Angela L. Duckworth et al., "Grit: Perseverance and Passion for Long-Term Goals," *Journal of Personality and Social Psychology* 92, no. 6 (2007): 1087. For a summary of grit as it relates to resilience, see John Grych, Sherry Hamby, and Victoria Banyard, "The Resilience Portfolio Model: Understanding Healthy Adaptation in Victims of Violence," *Psychology of Violence* 5, no. 4 (2015): 343. See also Angela Lee Duckworth, Tracy A. Steen, and Martin E. P. Seligman, "Positive Psychology in Clinical Practice," *Annual Review of Clinical Psychology* 1 (2005): 629–51; Sherry Hamby, John Grych, and Victoria Banyard, "Resilience Portfolios and Poly-Strengths: Identifying Protective Factors Associated with Thriving after Adversity," *Psychology of Violence* 8, no. 2 (2018): 172.

28 **learnable strengths**: Hamby, Grych, and Banyard, "Resilience Portfolios and Poly-Strengths," 172.

Notes

28 **optimism**: For a summary, see Gang Wu et al., "Understanding Resilience," *Frontiers in Behavioral Neuroscience* 7 (2013): 10; Lee et al., "Resilience: A Meta-Analytic Approach," 269–79.

28 **gratitude**: For a summary, see Grych, Hamby, and Banyard, "Resilience Portfolio Model," 343.

28 **exercise**: For a summary, see Wu et al., "Understanding Resilience," 10.

28 **active coping**: For an overview, see Wu et al., "Understanding Resilience," 10; for an example, see Ovidiu Popa-Velea et al., "Resilience and Active Coping Style: Effects on the Self-Reported Quality of Life in Cancer Patients," *International Journal of Psychiatry in Medicine* 52, no. 2 (2017): 124–36.

28 **applying several different practices**: Grych, Hamby, and Banyard, "Resilience Portfolio Model," 343.

29 **especially insects and birds**: Resilience Alliance, "Holling Fund," accessed September 9, 2024, https://www.resalliance.org/index.php/hollingfund.

29 **everyone use his prized moniker**: Lance Gunderson, "The Passing of a Polymath: In Memory of Buzz Holling," News, Resilience Alliance, accessed September 9, 2024, https://www.resalliance.org/news/51.

29 **followed by a PhD**: Stephen R. Carpenter and Garry D. Peterson, "C. S. 'Buzz' Holling, 6 December 1930–16 August 2019," Obituaries, *Nature Sustainability* 2 (2019): 997–98.

30 **couldn't spot their prey**: Stockholm Resilience Centre, "Buzz Holling Resilience Dynamics," posted on YouTube, January 23, 2009, https://www.youtube.com/watch?v=FrNWUOmOHRs.

30 **state of delicate balance**: Italics added for emphasis. Crawford Stanley Holling and Lance H. Gunderson, *Panarchy: Understanding Transformations in Human and Natural Systems* (Washington, DC: Island Press, 2002), 12.

30 **inherent feature of the natural world**: Tomas Weber, "The Strange Story of the Man behind the Popularization of 'Resilience Thinking,'" *Boston Globe*, March 2, 2023.

30 **how well they could adapt**: Stockholm Resilience Centre, "Buzz Holling Resilience Dynamics."

30 **the capacity of a system**: Specifically, Holling defined resilience as "the persistence of systems and . . . their ability to absorb change and disturbance and still maintain [functioning]" (Crawford S. Holling, "Resilience and Stability of Ecological Systems," *Annual Review of Ecology and Systematics* 4, no. 1 (1973): 1–23).

30 **strengthen our capacity to adapt?**: Weber, "Strange Story."

31 **the Latin *resilire***: Shae-Leigh Cynthia Vella and Nagesh B. Pai, "A Theoretical Review of Psychological Resilience: Defining Resilience and Resilience Research over the Decades," *Archives of Medicine and Health Sciences* 7, no. 2 (2019): 233–39i.

31 **"spring back"**: Latdict, s.v., "resilio, resilire, resilui," accessed September 9, 2024, https://latin-dictionary.net/definition/33432/resilio-resilire-resilui.

31 **Christianity**: In the book of Job, "Job is a prosperous man of outstanding piety. Satan acts as an agent provocateur to test whether or not Job's piety is rooted merely in his prosperity. But faced with the appalling loss of his possessions, his children, and finally his own health, Job still refuses to curse God" (Editors of Encyclopedia, "The Book of Job," *Encyclopedia Britannica*, last updated July 31, 2024, https://www.britannica.com/topic/The-Book-of-Job). Other relevant Bible verses: "Weeping may endure for a night. But joy comes in the morning" (Psalm 30:5); "Blessed is the man who remains steadfast under trial, for when he has stood

the test he will receive the crown of life, which God has promised to those who love him" (James 1:12); "I have said these things to you, that in me you may have peace. In the world you will have tribulation. But take heart; I have overcome the world" (John 16:33).

31 **Exodus**: P. J. Clyde Randall and Ed Winn, *The Exodus* (Pittsburgh, PA: Peoples Printing, 1919).

31 **Islam**: "Be sure we shall test you with something of fear and hunger, some loss in goods, lives, and the fruits of your toil. But give glad tidings to those who patiently persevere. Those who say, when afflicted with calamity, 'To Allah we belong, and to Him is our return.' They are those on whom descend blessings from their Lord, and mercy. They are the ones who receive guidance" (Koran 2:155–57).

31 **Buddhism**: Editors of Encyclopedia, "dukkha," *Encyclopedia Britannica*, November 26, 2003, https://www.britannica.com/topic/dukkha; Donald S. Lopez, "Four Noble Truths," *Encyclopedia Britannica*, August 11, 2023, https://www.britannica.com/topic/Four-Noble-Truths.

31 **Hinduism**: "Let me not beg for the stilling of my pain but for the heart to conquer it," quote from Rabindranath Tagore (1861–1941), Indian Hindu mystic philosopher. "A central life's work is to become detached from overinvolvement in the world" (Sarah M. Whitman, "Pain and Suffering as Viewed by the Hindu Religion," *Journal of Pain* 8, no. 8 (2007): 607–13.

31 **Stoicism**: "You have power over your mind—not outside events. Realize this, and you will find strength" (Marcus Aurelius, *Meditations*, 8.47).

31 **why some children successfully rebounded**: Lynne Michael Blum and R. W. Blum, "Resilience in Adolescence," *Adolescent Health: Understanding and Preventing Risk Behaviors* 1 (2009): 51–76.

31 **began publishing their results**: Emmy Werner et al., "Reproductive and Environmental Casualties: A Report on the 10-Year Follow-Up of the Children of the Kauai Pregnancy Study," *Pediatrics* 42, no. 1 (1968): 112–27.

31 **some children *were* uniquely able to bounce back**: They found that about one-third of their sample fell into this category, and of that group, one-third showed resilient outcomes (Emmy Werner, "Risk, Resilience, and Recovery," *Reclaiming Children and Youth* 21, no. 1 [2012]: 18).

31 **didn't start using the term**: Emmy E. Werner and Ruth S. Smith, "A Report from the Kauai Longitudinal Study," *Journal of the American Academy of Child Psychiatry* 18, no. 2 (1979): 292–306.

31 **some psychologists noting the dangers**: Suniya S. Luthar and Dante Cicchetti, "The Construct of Resilience: Implications for Interventions and Social Policies," *Development and Psychopathology* 12, no. 4 (2000): 857–85, at 862.

32 **researchers started studying the *process***: Ann S. Masten, "Resilience in Developing Systems: Progress and Promise as the Fourth Wave Rises," *Development and Psychopathology* 19, no. 3 (2007): 921–30.

32 **self-help books like *The Resilience Factor***: Karen Reivich and Andrew Shatté, *The Resilience Factor: 7 Essential Skills for Overcoming Life's Inevitable Obstacles* (New York: Broadway Books, 2002).

32 **economic inequality**: Melissa S. Abelev, "Advancing Out of Poverty: Social Class Worldview and Its Relation to Resilience," *Journal of Adolescent Research* 24, no. 1 (2009): 114–41.

Notes

32 **natural disasters**: Richard Haigh and Dilanthi Amaratunga, "An Integrative Review of the Built Environment Discipline's Role in the Development of Society's Resilience to Disasters," *International Journal of Disaster Resilience in the Built Environment* 1, no. 1 (2010): 11–24.

32 **risk management**: Denis Smith and Moira Fischbacher, "The Changing Nature of Risk and Risk Management: The Challenge of Borders, Uncertainty and Resilience," *Risk Management* 11 (2009): 1–12.

32 **business**: Happy Paul and Pooja Garg, "Elevating Organizational Consequences through Employee Resilience," National Conference on Emerging Challenges for Sustainable Business, 2012.

32 **education**: Jean E. Brooks, "Strengthening Resilience in Children and Youths: Maximizing Opportunities through the Schools," *Children & Schools* 28, no. 2 (2006): 69–76.

32 **therapy**: Froma Fam Walsh, "Family Resilience: A Developmental Systems Framework," *European Journal of Developmental Psychology* 13, no. 3 (2016): 313–24.

32 **healthcare**: Nebil Achour and Andrew D. F. Price, "Resilience Strategies of Healthcare Facilities: Present and Future," *International Journal of Disaster Resilience in the Built Environment* 1, no. 3 (2010): 264–76.

32 **engineering**: Jean Pariès and John Wreathall, *Resilience Engineering in Practice: A Guidebook* (Boca Raton, FL: CRC Press, 2017).

32 **infrastructure**: FEMA, "Building Resilient Infrastructure and Communities," US Department of Homeland Security, accessed September 9, 2024, https://www.fema.gov/grants/mitigation/building-resilient-infrastructure-communities#:~:text=Building%20Resilient%20Infrastructure%20and%20Communities%20(BRIC)%20supports%20states%2C%20local,from%20disasters%20and%20natural%20hazards.

32 *The Art of Resilience*: Ross Edgley, *The Art of Resilience* (New York: HarperCollins, 2020).

32 **Michelle Obama**: Apollo Theater (@apollotheater), "#FLOTUS shared sage advice for the girls gathered at today's #GlamourforEDU #62MillionGirls event!," Instagram, September 29, 2015, https://www.instagram.com/p/8OPwbGSRRS/?utm_source=ig_embed; see also Michelle Obama, *Becoming* (New York: Crown, 2021).

32 **Jennifer Aniston**: Jennifer Aniston Quotes, Imdb.com. www.imdb.com/name/nm0000098/quotes/.

32 **Elon Musk**: Ronald Brakels, "Elon Musk Delivers Surprise 'Resilience Day' Presentation," *SolarQuotes Blog*, April 1, 2022, https://www.solarquotes.com.au/blog/elon-musk-resilience-day/.

32 **companies send employees to resilience training**: Ellen Barry, "Workplace Wellness Programs Have Little Benefit, Study Finds," *New York Times*, January 15, 2024.

32 **"the most underrated and powerful skill"**: David Villa, "The Power of Resilience, and How to Develop It," *Forbes*, March 26, 2021, https://www.forbes.com/sites/forbesagencycouncil/2021/03/26/the-power-of-resilience-and-how-to-develop-it/?sh=1d76c9dc1f2a.

33 **focused on the idea of strength**: This is aligned with the American Psychological Association's definition of resilience, where bouncing back is central to the concept (Bruce W. Smith et al., "The Brief Resilience Scale: Assessing the Ability to Bounce Back," *International Journal of Behavioral Medicine* 15 (2008): 194–200; Charles S. Carver, "Resilience and Thriving: Issues, Models, and Linkages," *Journal of Social Issues* 54, no. 2 (1998): 245–66).

34 **resilience during COVID lockdowns**: Anna Panzeri et al., "Factors Impacting Resilience as a Result of Exposure to COVID-19: The Ecological Resilience Model," *PLOS One* 16, no. 8 (2021): e0256241.

Notes

34   **Several key literature reviews**: A few excellent reviews of this debate include Antonella Sisto et al., "Towards a Transversal Definition of Psychological Resilience: A Literature Review," *Medicina* 55, no. 11 (2019): 745; Gemma Aburn, Merryn Gott, and Karen Hoare, "What Is Resilience? An Integrative Review of the Empirical Literature," *Journal of Advanced Nursing* 72, no. 5 (2016): 980–1000; Richard Reid and Linda Courtenay Botterill, "The Multiple Meanings of 'Resilience': An Overview of the Literature," *Australian Journal of Public Administration* 72, no. 1 (2013): 31–40. Then, two great articles do a great job of narratively summarizing this debate: Steven M. Southwick et al., "Resilience Definitions, Theory, and Challenges: Interdisciplinary Perspectives," *European Journal of Psychotraumatology* 5, no. 1 (2014): 25338; Christy A. Denckla et al., "Psychological Resilience: An Update on Definitions, A Critical Appraisal, and Research Recommendations," *European Journal of Psychotraumatology* 11, no. 1 (2020): 1822064.

34   **the fifty-two most highly cited**: Most articles on the science of resilience mention the lack of consensus in defining resilience. Two examples that really make this debate come alive: Southwick, "Resilience Definitions, Theory, and Challenges," 25338; Denckla et al., "Psychological Resilience," 1822064.

35   **"better than expected"**: Suniya S. Luthar, Dante Cicchetti, and Bronwyn Becker, "The Construct of Resilience: A Critical Evaluation and Guidelines for Future Work," *Child Development* 71, no. 3 (2000): 543–62.

35   **"better than average"**: Marisa E. Hilliard, Michael A. Harris, and Jill Weissberg-Benchell, "Diabetes Resilience: A Model of Risk and Protection in Type 1 Diabetes," *Current Diabetes Reports* 12 (2012): 739–48.

35   **per resilience researchers themselves**: Ann S. Masten and Marie-Gabrielle J. Reed, "Resilience in Development," *Handbook of Positive Psychology* 74 (2002): 88.

35   **resilient people**: As measured by the Brief Resilience Scale, one of the three most widely respected scales for measuring resilience (Bruce W. Smith et al., "The Brief Resilience Scale: Assessing the Ability to Bounce Back," *International Journal of Behavioral Medicine* 15 [2008]: 194–200; see also Isyaku Salisu and Norashida Hashim, "A Critical Review of Scales Used in Resilience Research," *IOSR Journal of Business and Management* 19, no. 4 [2017]: 23–33).

36   **helping us maintain functioning**: Daniel J. Brown, Mustafa Sarkar, and Karen Howells, "Growth, Resilience, and Thriving: A Jangle Fallacy?," in *Growth Following Adversity in Sport: A Mechanism to Positive Change*, ed. Ross Wadey, Melissa Day, and Karen Howells (New York: Routledge, 2020), 59–72.

36   **drawing on resilience in hard times**: Hamideh Mahdiani and Michael Ungar, "The Dark Side of Resilience," *Adversity and Resilience Science* 2.3 (2021): 147–55.

37   **"studies of studies"**: that is, meta-analyses and systematic reviews.

37   **positive thinking and social support**: Isabella Helmreich et al., "Psychological Interventions for Resilience Enhancement in Adults," *Cochrane Database of Systematic Reviews* 2017, no. 2 (2017).

37   **reappraisal and acceptance**: Adam J. Vanhove et al., "Can Resilience Be Developed at Work? A Meta-Analytic Review of Resilience-Building Programme Effectiveness," *Journal of Occupational and Organizational Psychology* 89, no. 2 (2016): 278–307.

37   **but only slightly**: Jenny J. W. Liu et al., "Comprehensive Meta-Analysis of Resilience Interventions," *Clinical Psychology Review* 82 (2020): 101919; Richta C. IJntema, Yvonne D.

Burger, and Wilmar B. Schaufeli, "Reviewing the Labyrinth of Psychological Resilience: Establishing Criteria for Resilience-Building Programs," *Consulting Psychology Journal: Practice and Research* 71, no. 4 (2019): 288; Sadhbh Joyce et al., "Road to Resilience: A Systematic Review and Meta-Analysis of Resilience Training Programmes and Interventions," *British Medical Journal Open* 8, no. 6 (2018); Helmreich et al., "Psychological Interventions for Resilience Enhancement in Adults."

37 **don't improve resilience**: One meta-analysis of resilience not predicting mental health or well-being: Erik van der Meulen et al., "Longitudinal Associations of Psychological Resilience with Mental Health and Functioning among Military Personnel: A Meta-Analysis of Prospective Studies," *Social Science & Medicine* 255 (2020): 112814.

37 **they actually harmed mental health**: William J. Fleming, "Employee Well-Being Outcomes from Individual-Level Mental Health Interventions: Cross-Sectional Evidence from the United Kingdom," *Industrial Relations Journal* 55, no. 2 (2024): 162–82.

37 **"arrogant and victim-blaming"**: Brian Bethune, "When It Comes to Resilience, the Self-Help Industry Has It All Wrong," Maclean's, May 23, 2019, https://macleans.ca/society/when-it-comes-to-resilience-the-self-help-industry-has-it-all-wrong/.

37 **shifts our focus away**: Weber, "Strange Story."

38 **gaslighting**: Merriam-Webster defines the term *gaslighting* as the "psychological manipulation of a person usually over an extended period of time that causes the victim to question the validity of their own thoughts, perception of reality, or memories and typically leads to confusion, loss of confidence and self-esteem, uncertainty of one's emotional or mental stability, and a dependency on the perpetrator" (Merriam-Webster, s.v., "gaslighting," accessed September 9, 2024, https://www.merriam-webster.com/dictionary/gaslighting).

38 **several factors outside of our control**: Belinda Bruwer et al., "Psychometric Properties of the Multidimensional Scale of Perceived Social Support in Youth," *Comprehensive Psychiatry* 49, no. 2 (2008): 195–201.

38 **DNA**: For a summary of the genetic factors at play in resilience, see Gang Wu et al., "Understanding Resilience," *Frontiers in Behavioral Neuroscience* 7 (2013): 10. For a summary of epigenetic factors in resilience, see Demelza Smeeth et al., "The Role of Epigenetics in Psychological Resilience," *Lancet Psychiatry* 8, no. 7 (2021): 620–29.

38 **nervous system**: Haoran Liu et al., "Biological and Psychological Perspectives of Resilience: Is It Possible to Improve Stress Resistance?," *Frontiers in Human Neuroscience* 12 (2018): 326.

38 **personality**: Helen Herrman et al., "What Is Resilience?," *Canadian Journal of Psychiatry* 56, no. 5 (2011): 258–65.

38 **temperament**: Craig A. Olsson et al., "Adolescent Resilience: A Concept Analysis," *Journal of Adolescence* 26, no. 1 (2003): 1–11. This includes things like positive emotions (Bart P. F. Rutten et al., "Resilience in Mental Health: Linking Psychological and Neurobiological Perspectives," *Acta Psychiatrica Scandinavica* 128, no. 1 [2013]: 3–20); Stevan E. Hobfoll, Natalie R. Stevens, and Alyson K. Zalta, "Expanding the Science of Resilience: Conserving Resources in the Aid of Adaptation," *Psychological Inquiry* 26, no. 2 (2015): 174–80.

38 **early childhood experiences**: Katie A. McLaughlin et al., "Childhood Adversity, Adult Stressful Life Events, and Risk of Past-Year Psychiatric Disorder: A Test of the Stress Sensitization Hypothesis in a Population-Based Sample of Adults," *Psychological Medicine* 40, no. 10 (2010): 1647–58.

38 **later life events**: Adverse childhood experiences cause resilience-diminishing changes in our brains and stress-response systems. See Wu et al., "Understanding Resilience," 10. Early childhood adversity also has been shown to diminish grit: Shannon Cheung, Chien-Chung Huang, and Congcong Zhang, "Passion and Persistence: Investigating the Relationship between Adverse Childhood Experiences and Grit in College Students in China," *Frontiers in Psychology* 12 (2021): 642956.
- People with low socioeconomic status have double the rates of psychiatric illness as compared to their high socioeconomic status counterparts (Charles E. Holzer, B. M. Shea, J. W. Swanson, P. J. Leaf et al., "The Increased Risk for Specific Psychiatric Disorders among Persons of Low Socioeconomic Status," *American Journal of Social Psychiatry* 6, no. 4 [1986]: 259–71).
    - For example, one study of police recruits found that those with a history of childhood trauma showed higher levels of salivary stress hormones than recruits without that history (Christian Otte et al., "Association between Childhood Trauma and Catecholamine Response to Psychological Stress in Police Academy Recruits," *Biological Psychiatry* 57, no. 1 [2005]: 27–32).
- What's more, our experience with stress, adversity, and trauma over our lifetime can also make us less resilient. See Chris R. Brewin, Bernice Andrews, and John D. Valentine, "Meta-Analysis of Risk Factors for Posttraumatic Stress Disorder in Trauma-Exposed Adults," *Journal of Consulting and Clinical Psychology* 68, no. 5 [2000]: 748; Mark D. Seery, E. Alison Holman, and Roxane Cohen Silver, "Whatever Does Not Kill Us: Cumulative Lifetime Adversity, Vulnerability, and Resilience," *Journal of Personality and Social Psychology* 99, no. 6 [2010]: 1025; Paula S. Nurius, Edwina Uehara, and Douglas F. Zatzick, "Intersection of Stress, Social Disadvantage, and Life Course Processes: Reframing Trauma and Mental Health," *American Journal of Psychiatric Rehabilitation* 16, no. 2 [2013]: 91–114; George A. Bonanno et al., "What Predicts Psychological Resilience after Disaster? The Role of Demographics, Resources, and Life Stress," *Journal of Consulting and Clinical Psychology* 75, no. 5 [2007]: 671.

39 **published a book**: Friedrich Wilhelm Nietzsche, *The Twilight of the Idols* (Crows Nest, Australia: Allen & Unwin, 1974).

39 **during a morning stroll**: T. C. Williams, "The Last Years of Friedrich Nietzsche: The Horse Incident and Its Aftermath," *Medium*, April 23, 2022, https://medium.com/@T.C.Williams/the-last-years-of-friedrich-nietzsche-the-horse-incident-and-its-aftermath-97c9559fdcf#:~:text=On%20the%20morning%20of%20January,to%20the%20ground%20in%20tears.

39 **"never emerged again"**: Tim Brinkhof, "Did Friedrich Nietzsche's Own Philosophy Drive Him Insane?" Big Think, October 21, 2023, https://bigthink.com/high-culture/friedrich-nietzsche-insanity/. It's also worth noting that the cause of Nietzche's death is still unknown, but there is general consensus that he died from either a form of dementia or syphilis. Shrewd students of history might take issue with my implication that Nietzsche's descent into madness was the sole result of reaching the limits of resilience. To them I say, fair point! But this story was too good not to include with a "spirit of the law" caveat.

39 **"lacks robust empirical evidence"**: Frank J. Infurna and Eranda Jayawickreme, "Fixing the Growth Illusion: New Directions for Research in Resilience and Posttraumatic Growth," *Current Directions in Psychological Science* 28, no. 2 (2019): 152–58. These authors also argue that the appeal of the idea of adversity building strength may be what led scientists to claim a little too hastily that growth is a typical response to adversity (it's not). See also Seery, Holman, and Cohen Silver, "Whatever Does Not Kill Us," 1025.

Notes

39 **most limited resilience resources**: That is, greater social, educational, and intellectual disadvantage (Brewin, Andrews, and Valentine, "Meta-Analysis of Risk Factors for Posttraumatic Stress Disorder in Trauma-Exposed Adults," 748).

39 **experienced the most stress**: Bruwer et al., "Psychometric Properties of the Multidimensional Scale of Perceived Social Support in Youth," 195–201.

39 **exhaustible resource that dwindles**: David Fletcher and Mustafa Sarkar, "Psychological Resilience," *European psychologist* (2013); Denckla et al., "Psychological Resilience," 1822064; Luthar, Cicchetti, and Becker, "Construct of Resilience," 543–62; Seery, Holman, and Cohen Silver, "Whatever Does Not Kill Us," 1025.

39 **ongoing stress tends to *deplete* it**: Cristina A. Fernandez et al., "Assessing the Relationship between Psychosocial Stressors and Psychiatric Resilience among Chilean Disaster Survivors," *British Journal of Psychiatry* 217, no. 5 (2020): 630–37.

40 **even for minor stressors**: Scott M. Monroe and Kate L. Harkness, "Life Stress, the 'Kindling' Hypothesis, and the Recurrence of Depression: Considerations from a Life Stress Perspective," *Psychological Review* 112, no. 2 (2005): 417. In fact, minor stress over time might be even more depleting than trauma. See Jose Manuel Rodriguez-Llanes, Femke Vos, and Debarati Guha-Sapir, "Measuring Psychological Resilience to Disasters: Are Evidence-Based Indicators an Achievable Goal?," *Environmental Health* 12, no. 1 (2013): 1–10; Kenneth E. Miller et al., "Daily Stressors, War Experiences, and Mental Health in Afghanistan," *Transcultural Psychiatry* 45, no. 4 (2008): 611–38.

40 **across more than one life domain**: A helpful concept is allostatic load. It refers to the wear-and-tear our bodies experience amid repeated exposure to stress. (For more, see Nurius, Uehara, and Zatzick, "Intersection of Stress, Social Disadvantage, and Life Course Processes," 91–114.) For a great explanation of how stress drains resilience, see Carmelina Lawton Smith, "Coaching for Leadership Resilience: An Integrated Approach," *International Coaching Psychology Review* 12, no. 1 (2017): 6–23.

40 **lowered participants' resilience over time**: Zhang Zhao, Rui Wan, and JingDan Ma, "Social Change and Birth Cohorts Decreased Resilience among College Students in China: A Cross-Temporal Meta-Analysis, 2007–2020," *Personality and Individual Differences* 196 (2022): 111716.

40 **watched a distressing movie**: Mark Muraven, Dianne M. Tice, and Roy F. Baumeister, "Self-Control as a Limited Resource: Regulatory Depletion Patterns," *Journal of Personality and Social Psychology* 74, no. 3 (1998): 774.

40 **it will fail completely**: This doesn't mean that when our resilience reserves are running low that they can't be replenished, but it does mean that recovery will take time—often times we don't have (Stevan E. Hobfoll et al., "The Limits of Resilience: Distress Following Chronic Political Violence among Palestinians," *Social Science & Medicine* 72, no. 8 (2011): 1400–1408).

40 **members of marginalized groups**: The goal of resilience can be especially problematic for marginalized people who experience sustained challenges that others don't. For example, researcher Stevan Hobfoll notes that the availability of social support often depends on power or status, and researcher Kristina Diprose points out that when our culture celebrates overcoming the odds, resource-constrained people who are constantly striving are more likely to develop learned helplessness than resilience (Stevan E. Hobfoll and Jennifer D. Wells, "Conservation of Resources, Stress, and Aging: Why Do Some Slide and Some Spring?," in *Handbook of Aging and Mental Health: An Integrative Approach*, ed. Jacob Lomranz [Boston, MA:

Springer US, 1998], 121–34; Kristina Diprose, "Resilience Is Futile," *Soundings*, no. 58 [2015]: 44–56). And while in the minority of opinions, it is worth noting that some scholars criticize the very concept of resilience as "a tool for social containment in ways that align with ableist, sexist, colonial, and neoliberal interests" (Emily Hutcheon and Bonnie Lashewicz, "Theorizing Resilience: Critiquing and Unbounding a Marginalizing Concept," *Disability & Society* 29, no. 9 [2014]: 1383–97; see also Hamideh Mahdiani and Michael Ungar, "The Dark Side of Resilience," *Adversity and Resilience Science* 2, no. 3 [2021]: 147–55).

40 **because of his ethnicity**: Simran Jeet Singh, "'Resilient' Isn't the Compliment You think It Is," *Harvard Business Review*, March 22, 2023, https://hbr.org/2023/03/resilient-isnt-the-compliment-you-think-it-is.

41 **mental illness**: Carmen Valiente et al., "A Symptom-Based Definition of Resilience in Times of Pandemics: Patterns of Psychological Responses over Time and Their Predictors," *European Journal of Psychotraumatology* 12, no. 1 (2021): 1871555; Paula P. Schnurr, Matthew J. Friedman, and Stanley D. Rosenberg, "Premilitary MMPI Scores as Predictors of Combat-Related PTSD Symptoms," *American Journal of Psychiatry* 150, no. 3 (1993): 479–83; Brewin, Andrews, and Valentine, "Meta-Analysis of Risk Factors for Posttraumatic Stress Disorder in Trauma-Exposed Adults," 748.

41 **discrimination**: For example, see Gene H. Brody et al., "Perceived Discrimination among African American Adolescents and Allostatic Load: A Longitudinal Analysis with Buffering Effects," *Child Development* 85, no. 3 (2014): 989–1002. Some studies have suggested that African Americans experience greater levels of stressors after controlling for socioeconomic status—and discrimination is likely a large part of why (for an example, see Ronald C. Kessler, Kristin D. Mickelson, and David R. Williams, "The Prevalence, Distribution, and Mental Health Correlates of Perceived Discrimination in the United States," *Journal of Health and Social Behavior* 40, no. 3 [1999]: 208–30). For more information, see Nurius, Uehara, and Zatzick, "Intersection of Stress, Social Disadvantage, and Life Course Processes," 91–114.

41 **chronic illness and disability**: See Constance Hammen, "Stress and Depression," *Annual Review of Clinical Psychology* 1 (2005): 293–319. On chronic illness, see Bonanno et al., "What Predicts Psychological Resilience after Disaster?," 671. On chronic disease, see Ljiljana Trtica Majnarić et al., "Low Psychological Resilience in Older Individuals: An Association with Increased Inflammation, Oxidative Stress and the Presence of Chronic Medical Conditions," *International Journal of Molecular Sciences* 22, no. 16 (2021): 8970.

41 **being female**: G. Ö. K. Ayşe and Esin Yilmaz Koğar, "A Meta-Analysis Study on Gender Differences in Psychological Resilience Levels," *Kıbrıs Türk Psikiyatri ve Psikoloji Dergisi* 3, no. 2 (2021): 132–43.

41 **"It does not seem worth it"**: Brinkhof, "Did Friedrich Nietzsche's Own Philosophy Drive Him Insane?"

41 **we all have our limits**: Michael Rutter, "Resilience: Some Conceptual Considerations," in *Social Work*, ed. Viviene E. Cree and Trish McCulloch, 2nd ed. (London: Routledge, 2023): 122–127.

41 **one "right way"**: Uisce Jordan, "A Visual Investigation of Resilience from the Perspective of Social Work Students in the UK," *Social Work Education*, 43, no. 4 (2024): 1-15; John Harvey and Paul H. Delfabbro, "Psychological Resilience in Disadvantaged Youth: A Critical Overview," *Australian Psychologist* 39, no. 1 (2004): 3–13.

Notes

Chapter 3: Hitting Our Resilience Ceiling

45 **not an inexhaustible resource**: As Mark Seery points out, just because we stayed stable or recovered from one thing doesn't give us a greater future capacity for future hard things (Mark D. Seery, E. Alison Holman, and Roxane Cohen Silver, "Whatever Does Not Kill Us: Cumulative Lifetime Adversity, Vulnerability, and Resilience," *Journal of Personality and Social Psychology* 99, no. 6 [2010]: 1025; see also Elizabeth Harrison, "Bouncing Back? Recession, Resilience and Everyday Lives," *Critical Social Policy* 33, no. 1 [2013]: 97–113; E. Anne Bardoel and Robert Drago, "Acceptance and Strategic Resilience: An Application of Conservation of Resources Theory," *Group & Organization Management* 46, no. 4 [2021]: 657–91; Michael Neenan, "Resilience Coaching," *Coaching for Rational Living: Theory, Techniques and Applications*, ed. Michael E. Bernard and Oana A. David [Cham, Switzerland: Springer, 2018], 247–67; Paula S. Nurius, Edwina Uehara, and Douglas F. Zatzick, "Intersection of Stress, Social Disadvantage, and Life Course Processes: Reframing Trauma and Mental Health," *American Journal of Psychiatric Rehabilitation* 16, no. 2 [2013]: 91–114; Monique F. Crane and Ben J. Searle, "Building Resilience through Exposure to Stressors: The Effects of Challenges versus Hindrances," *Journal of Occupational Health Psychology* 21, no. 4 (2016): 468; Stevan E. Hobfoll, Natalie R. Stevens, and Alyson K. Zalta, "Expanding the Science of Resilience: Conserving Resources in the Aid of Adaptation," *Psychological Inquiry* 26, no. 2 (2015): 174–80).

46 **"biologically implausible"**: Michael Rutter, "Resilience: Some Conceptual Considerations," in *Social Work*, ed. Viviene E. Cree and Trish McCulloch, 2nd ed. (London: Routledge, 2023): 122–27.

46 **less successful coping**: Hamideh Mahdiani and Michael Ungar, "The Dark Side of Resilience," *Adversity and Resilience Science* 2, no. 3 (2021): 147–55; Stephanie MacLeod et al., "The Impact of Resilience among Older Adults," *Geriatric Nursing* 37, no. 4 (2016): 266–72.

46 **more distress**: Stevan E. Hobfoll, "Conservation of Resources: A New Attempt at Conceptualizing Stress," *American Psychologist* 44, no. 3 (1989): 513.

46 **discouragement, and shame**: Janet Polivy and C. Peter Herman, "The False-Hope Syndrome: Unfulfilled Expectations of Self-Change," *Current Directions in Psychological Science* 9, no. 4 (2000): 128–31.

46 **grit, a close cousin of resilience**: Angela L. Duckworth et al., "Grit: Perseverance and Passion for Long-Term Goals," *Journal of Personality and Social Psychology* 92, no. 6 (2007): 1087.

46 **myriad positive outcomes**: For example, Marcus Credé, Michael C. Tynan, and Peter D. Harms, "Much Ado about Grit: A Meta-Analytic Synthesis of the Grit Literature," *Journal of Personality and social Psychology* 113, no. 3 (2017): 492.

47 **better path for our well-being**: Gale M. Lucas et al., "When the Going Gets Tough: Grit Predicts Costly Perseverance," *Journal of Research in Personality* 59 (2015): 15–22.

47 **burnout, the emotional exhaustion**: Christina Maslach, Susan E. Jackson, and Michael P. Leiter, *Maslach Burnout Inventory* (Lanham, MD: Scarecrow Education, 1997).

47 **too much work with too little impact**: Personal communication.

47 **the more vulnerable we become to the stressors**: Seery, Holman, and Cohen Silver, "Whatever Does Not Kill Us," 1025.

48 **we tend not to be self-aware about our resilience**: Peta Sigley-Taylor, Tan-Chyuan Chin, and Dianne A. Vella-Brodrick, "Do Subjective and Objective Resilience Measures Assess Unique Aspects and What Is Their Relationship to Adolescent Well-Being?," *Psychology in the*

Notes

Schools 58, no. 7 (2021): 1320–44; George A. Bonanno, Maren Westphal, and Anthony D. Mancini, "Resilience to Loss and Potential Trauma," *Annual Review of Clinical Psychology* 7 (2011): 511–35.

Similarly, one common way of protecting ourselves from psychological pain is to inflate our ego through something psychologists call "self-enhancement," or the tendency to make overly positive or inaccurate self-serving biases (George A. Bonanno, Courtney Rennicke, and Sharon Dekel, "Self-Enhancement among High-Exposure Survivors of the September 11th Terrorist Attack: Resilience or Social Maladjustment?," *Journal of Personality and Social Psychology* 88, no. 6 [2005]: 984). And while this helps us cope with bad things, self-enhancers, on average, are seen by others as less honest, and less well-adjusted, and comes with other social costs (Bonanno, Rennicke, and Dekel, "Self-Enhancement among High-Exposure Survivors of the September 11th Terrorist Attack: Resilience or Social Maladjustment?"; Shelley E. Taylor and David A. Armor, "Positive Illusions and Coping with Adversity," *Journal of Personality* 64, no. 4 [1996]: 873–98; George A. Bonanno et al., "Self-Enhancement as a Buffer against Extreme Adversity: Civil War in Bosnia and Traumatic Loss in the United States," *Personality and Social Psychology Bulletin* 28, no. 2 [2002]: 184–96; Mahdiani and Ungar, "Dark Side of Resilience," 147–55; Vera Hoorens, "The Social Consequences of Self-Enhancement and Self-Protection," in *Handbook of Self-Enhancement and Self-Protection*, ed. Mark D. Alicke and Constantine Sedikides [New York: Guildford Press, 2011], 235–57; Delroy L. Paulhus, "Interpersonal and Intrapsychic Adaptiveness of Trait Self-Enhancement: A Mixed Blessing?," *Journal of Personality and Social Psychology* 74, no. 5 [1998]: 1197).

48 **one survey revealed that 83 percent**: Everyday Health, "State of Health: Resilience," special report, Ohio State University, accessed September 10, 2024, https://images.agoramedia.com/everydayhealth/gcms/Everyday-Health-State-of-Health-Resilience.pdf.

48 **overrelying on resilience can be a source of fragility**: Mahdiani and Ungar, "Dark Side of Resilience," 147–55.

48 **Known as spoon theory**: Christine Miserandino, "The Spoon Theory," Lymphoma Action, accessed September 10, 2024, https://lymphoma-action.org.uk/sites/default/files/media/documents/2020-05/Spoon%20theory%20by%20Christine%20Miserandino.pdf.

49 **do less not more**: Charles S. Carver, "Resilience and Thriving: Issues, Models, and Linkages," *Journal of Social Issues* 54, no. 2 (1998): 245–66.

50 **substantial physical and psychological consequences**: Elizabeth I. Johnson and Joyce A. Arditti, "Risk and Resilience among Children with Incarcerated Parents: A Review and Critical Reframing," *Annual Review of Clinical Psychology* 19 (2023): 437–60.

51 **recent *Atlantic* article**: Sophie Gilbert, "How Did Healing Ourselves Get So Exhausting?," *Atlantic*, October 12, 2022, https://www.theatlantic.com/culture/archive/2022/10/goop-wellness-culture-self-care-parenting/671699/.

51 **habit of repressive coping**: George A. Bonanno and Jerome L. Singer, "Repressive Personality Style: Theoretical and Methodological Implications for Health and Pathology," in *Repression and Dissociation: Implications for Personality Theory, Psychopathology, and Health*, ed. Jerome L. Singer (Chicago: University of Chicago Press, 1990), 435–70; Michael Hock and Heinz Walter Krohne, "Coping with Threat and Memory for Ambiguous Information: Testing the Repressive Discontinuity Hypothesis," *Emotion* 4, no. 1 (2004): 65; Andrew J. Tomarken and Richard J. Davidson, "Frontal Brain Activation in Repressors and Nonrepressors," *Journal of Abnormal Psychology* 103, no. 2 (1994): 339.

Notes

51 **protect us from psychological pain in the short term**: Zahava Solomon, Roni Berger, and Karni Ginzburg, "Resilience of Israeli Body Handlers: Implications of Repressive Coping Style," *Traumatology* 13, no. 4 (2007): 64–74; see also Anthony D. Mancin and George A. Bonanno, "Predictors and Parameters of Resilience to Loss: Toward an Individual Differences Model," *Journal of Personality* 77, no. 6 (2009): 1805–32.

51 **costs us dearly in the long term**: Mancini and Bonanno, "Predictors and Parameters of Resilience to Loss," 1805–32.

51 **migraines and backaches**: Emmy Werner, "Risk, Resilience, and Recovery," *Reclaiming Children and Youth* 21, no. 1 (2012): 18.

51 **costly persistence**: Gale M. Lucas et al., "When the Going Gets Tough: Grit Predicts Costly Perseverance," *Journal of Research in Personality* 59 (2015): 15–22.

51 **downplay harsh realities**: William E. Rosa et al., "The Critical Need for a Meaning-Centered Team-Level Intervention to Address Healthcare Provider Distress Now," *International Journal of Environmental Research and Public Health* 19, no. 13 (2022): 7801.

51 **tolerate intolerable situations**: Tomas Chamorro-Premuzic and Derek Lusk, "The Dark Side of Resilience," *Harvard Business Review*, August 16, 2017, https://hbr.org/2017/08/the-dark-side-of-resilience.

51 **all of which rob us of agency**: Charles A. Ogunbode et al., "The Resilience Paradox: Flooding Experience, Coping and Climate Change Mitigation Intentions," *Climate Policy* 19, no. 6 (2019): 703–15; see also Kristina Diprose, "Resilience Is Futile," *Soundings*, no. 58 (2015): 44–56.

Chapter 4: The Second Skill Set

56 *Harvard Business Review*: Thomas Stackpole, "Inside IKEA's Digital Transformation," *Harvard Business Review*, June 4, 2021, https://hbr.org/2021/06/inside-ikeas-digital-transformation.

57 *HR Digest*: Jane Harper, "IKEA's Digital Transformation: How the Swedish Furniture Giant Is Adapting to the New Retail Landscape," *HR Digest*, June 2, 2023, https://www.thehrdigest.com/ikeas-digital-transformation-how-the-swedish-furniture-giant-is-adapting-to-the-new-retail-landscape/.

57 **McKinsey**: Retail Industry Leaders Association, "Retail Speaks: Seven Imperatives for the Industry," industry report, McKinsey & Company, accessed September 10, 2024, https://www.mckinsey.com/~/media/McKinsey/Industries/Retail/Our%20Insights/retail%20speaks%20seven%20imperatives%20for%20the%20industry/retail-speaks-full-report.pdf.

59 **This is the concept underpinning Kintsugi**: Kintsugi started with ceramics but was extended to substances like glass (Colin Marshall, "Kintsugi: The Centuries-Old Japanese Craft of Repairing Pottery with Gold & Finding Beauty in Broken Things," Open Culture, October 9, 2017, https://www.openculture.com/2017/10/kintsugi-the-centuries-old-japanese-craft-of-repairing-pottery-with-gold-finding-beauty-in-broken-things.html#google_vignette).

59 **under Shōgun Ashikaga Yoshimasa**: There is some disagreement around the factual basis of this story, but I am including it because it makes a powerful point, even if it's been somewhat fictionalized over the years (Sansho, "Kintsugi: Fact and Fiction," December 2, 2021, https://sansho.com/blogs/news/kintsugi-fact-and-fiction).

60 **addressed the United Negro College Fund**: "Remarks of Senator John F. Kennedy, Convocation of the United Negro College Fund, Indianapolis, Indiana, April 12, 1959," Papers of John F. Kennedy, Pre-presidential Papers, Senate Files, Box 902, United Negro College

Notes

Fund, Indianapolis, Indiana, John F. Kennedy Presidential Library, https://www.jfklibrary.org/archives/other-resources/john-f-kennedy-speeches/indianapolis-in-19590412.

60 **Chinese language and literature professor Victor Mair**: Victor H. Mair, "'Crisis' Does Not Equal 'Danger' Plus 'Opportunity,'" Pinyin.info, last revised September 2009, https://www.pinyin.info/chinese/crisis.html.

60 **Buzz Holling would likely agree**: Nicola Ross, "Calling Buzz Holling: An Exclusive Interview with the Father of Resilience Theory," *Alternatives Journal* 36, no. 2 (2010), https://go.gale.com/ps/i.do?p=AONE&u=googlescholar&id=GALE|A221746697&v=2.1&it=r&sid=AONE&asid=0cdfc32b.

60 **protect the status quo**: "Interview—Buzz Holling," special feature, *Alliance Magazine*, December 1, 2012, https://www.alliancemagazine.org/feature/interview-buzz-holling/.

60 **leaving us forever traumatized**: Ronald C. Kessler et al., "Trauma and PTSD in the WHO World Mental Health Surveys," *European Journal of Psychotraumatology* 8, sup. 5 (2017): 1353383; G. Perrotta, "Psychological Trauma: Definition, Clinical Contexts, Neural Correlations and Therapeutic Approaches," *Current Research in Psychiatry and Brain Disorders* 2019, no. 1 (2019).

61 **several ways individuals can emerge**: R. G. Tedeschi, C. Park, and L. G. Calhoun, *Post-Traumatic Growth: Theory and Research in the Aftermath of Crisis* (New York: Psychology Press, 2008).

61 **"post-traumatic growth"**: Lawrence G. Calhoun et al., "A Correlational Test of the Relationship between Posttraumatic Growth, Religion, and Cognitive Processing," *Journal of Traumatic Stress: Official Publication of The International Society for Traumatic Stress Studies* 13, no. 3 (2000): 521–27.

61 **more on endurance and recovery**: Sander L. Koole et al., "Becoming Who You Are: An Integrative Review of Self-Determination Theory and Personality Systems Interactions Theory," *Journal of Personality* 87, no. 1 (2019): 15–36.

62 **specific unmet psychological needs**: Adam Neufeld, Annik Mossière, and Greg Malin, "Basic Psychological Needs, More Than Mindfulness and Resilience, Relate to Medical Student Stress: A Case for Shifting the Focus of Wellness Curricula," *Medical Teacher* 42, no. 12 (2020): 1401–12.

62 **"primary nutrient of growth and wellness"**: Sander L. Koole et al., "Becoming Who You Are: An Integrative Review of Self-Determination Theory and Personality Systems Interactions Theory," *Journal of Personality* 87, no. 1 (2019): 15–36.

62 **specifically enhance need fulfillment**: Neufeld, Mossière, and Malin, "Basic Psychological Needs, More Than Mindfulness and Resilience, Relate to Medical Student Stress," 1401–12.

62 **shatterproof practices replenish them**: For a few interesting examples of this effect, see: Adam Neufeld and Greg Malin, "Exploring the Relationship between Medical Student Basic Psychological Need Satisfaction, Resilience, and Well-Being: A Quantitative Study," *BMC Medical Education* 19 (2019): 1–8; Dana Perlman et al., "A Path Analysis of Self-Determination and Resiliency for Consumers Living with Mental Illness," *Community Mental Health Journal* 54 (2018): 1239–44; Yuan Liu and Xiaoxing Huang, "Effects of Basic Psychological Needs on Resilience: A Human Agency Model," *Frontiers in Psychology* 12 (2021): 700035. For a review, see Maarten Vansteenkiste and Richard M. Ryan, "On Psychological Growth and Vulnerability: Basic Psychological Need Satisfaction and Need Frustration as a Unifying Principle," *Journal of Psychotherapy Integration* 23, no. 3 (2013): 263.

Notes

62 **energized, enthusiastic, and motivated**: Richard M. Ryan and Edward L. Deci, "From Ego Depletion to Vitality: Theory and Findings Concerning the Facilitation of Energy Available to the Self," *Social and Personality Psychology Compass* 2, no. 2 (2008): 702–17; see also Frank Martela, Cody R. DeHaan, and Richard M. Ryan, "On Enhancing and Diminishing Energy through Psychological Means: Research on Vitality and Depletion from Self-Determination Theory," in *Self-Regulation and Ego Control*, ed. Edward R. Hirt, Joshua J. Clarkson, and Lile Jia (London: Academic Press, 2016), 67–85.

63 **"the most reliable predictor"**: Here's a great article about this that you should read immediately: Marshall Goldsmith, "Don't Let Inertia Create Your Life," *Marshall Goldsmith* (blog), accessed September 10, 2024, https://marshallgoldsmith.com/articles/dont-let-inertia-create-your-life.

65 **pure, proactive transformation**: Netta Weinstein and Richard M. Ryan, "A Self-Determination Theory Approach to Understanding Stress Incursion and Responses," *Stress and Health* 27, no. 1 (2011): 4–17.

66 **On one wing**: Smithsonian Insider, "Dodo Bird a Resilient Island Survivor before the Arrival of Humans, Study Reveals," Dinosaurs & Fossils, Science & Nature, September 26, 2011, https://insider.si.edu/2011/09/dodo-bird-was-a-resilient-island-survivor-before-the-arrival-of-humans.

66 **blends into less snowy surroundings**: Patrik Karell, Kari Ahola, Teuvo Karstinen, Jari Valkama, and Jon E. Brommer, "Climate Change Drives Microevolution in a Wild Bird," *Nature Communications* 2, no. 208 (2011), https://doi.org/10.1038/ncomms1213.

66 **the tawny owl population isn't shrinking**: Susan Cosier, "Owl Populations Change Color as the World Warms," *Audubon*, April 2011, https://www.audubon.org/news/owl-populations-change-color-world-warms.

CHAPTER 5: STEP 1: PROBE YOUR PAIN

73 **As Marcus Aurelius once wrote**: Marcus Aurelius, *Meditations*, 8.47.

73 **big internal costs**: Jane M. Richards and James J. Gross, "Composure at Any Cost? The Cognitive Consequences of Emotion Suppression," *Personality and Social Psychology Bulletin* 25, no. 8 (1999): 1033–44; Holley S. Hodgins, Holly A. Yacko, and Ethan Gottlieb, "Autonomy and Nondefensiveness," *Motivation and Emotion* 30 (2006): 283–93, at 284.

73 **pulling our focus**: Steven C. Hayes et al., "Acceptance and Commitment Therapy: Model, Processes and Outcomes," *Behaviour Research and Therapy* 44, no. 1 (2006): 1–25; see also Richard M. Ryan, Bart Soenens, and Maarten Vansteenkiste, "Reflections on Self-Determination Theory as an Organizing Framework for Personality Psychology: Interfaces, Integrations, Issues, and Unfinished Business," *Journal of Personality* 87, no. 1 (2019): 115–45, at 126.

73 **draining our mental energy**: Frank Martela, Cody R. DeHaan, and Richard M. Ryan, "On Enhancing and Diminishing Energy through Psychological Means: Research on Vitality and Depletion from Self-Determination Theory," in *Self-Regulation and Ego Control*, ed. Edward R. Hirt, Joshua J. Clarkson, and Lile Jia (London: Academic Press, 2016), 67–85.

73 **messing with our mood**: Marilyn Mendolia and Robert E. Kleck, "Effects of Talking about a Stressful Event on Arousal: Does What We Talk about Make a Difference?," *Journal of Personality and Social Psychology* 64, no. 2 (1993): 283.

73 **increasing depression**: Ryan, Soenens, and Vansteenkiste, "Reflections on Self-Determination Theory as an Organizing Framework for Personality Psychology," 115–45.

Notes

73 ***more*** **negative emotions**: Robert Brockman et al., "Emotion Regulation Strategies in Daily Life: Mindfulness, Cognitive Reappraisal and Emotion Suppression," *Cognitive Behaviour Therapy* 46, no. 2 (2017): 91–113.

73 **"negativity rebounds"**: Tim Dalgleish et al., "Ironic Effects of Emotion Suppression When Recounting Distressing Memories," *Emotion* 9, no. 5 (2009): 744; see also Weinstein and Hodgins, "The Moderating Role of Autonomy," *Personality and Social Psychology Bulletin* 35, no. 3 (2009): 351–64.

73 **a trait centered on pushing through pain**: Graham Jones, Sheldon Hanton, and Declan Connaughton, "A Framework of Mental Toughness in the World's Best Performers," *Sport Psychologist* 21, no. 2 (2007): 243–64.

74 **is called toxic positivity**: Whitney Goodman, *Toxic Positivity: Keeping It Real in a World Obsessed with Being Happy* (New York: Penguin, 2022).

75 **aged seventeen to nineteen, it actually *increased* them**: Brockman et al., "Emotion Regulation Strategies in Daily Life," 91–113.

76 **physically immobilized, frozen, or completely detached**: Andrew Anthony, "Stephen Porges: 'Survivors Are Blamed Because They Don't Fight,'" *Guardian*, June 2, 2019, https://www.theguardian.com/society/2019/jun/02/stephen-porges-interview-survivors-are-blamed-polyvagal-theory-fight-flight-psychiatry-ace.

76 **freeze-or-faint**: Stephen W. Porges, *The Pocket Guide to the Polyvagal Theory: The Transformative Power of Feeling Safe* (New York: W. W. Norton, 2017), 54; see also Chris Eccleston and Geert Crombez, "Pain Demands Attention: A Cognitive–Affective Model of the Interruptive Function of Pain," *Psychological Bulletin* 125, no. 3 (1999): 356.

76 **total physical and emotional shutdown**: Margot Slade, "Stephen W. Porges, PhD: Q&A about Freezing, Fainting, and the 'Safe' Sounds of Music Therapy," Everyday Health, October 16, 2018, https://www.everydayhealth.com/wellness/united-states-of-stress/advisory-board/stephen-w-porges-phd-q-a/.

76 **is a confusing experience**: Slade, "Stephen W. Porges."

77 **it can actually reduce negative emotions**: For examples, Brockman et al., "Emotion Regulation Strategies in Daily Life," 91–113; Alessandro Grecucci et al., "Mindful Emotion Regulation: Exploring the Neurocognitive Mechanisms behind Mindfulness," *BioMed Research International* (2015): DOI: 10.1155/2015/670724.

77 **two days later**: Netta Weinstein and Holley S. Hodgins, "The Moderating Role of Autonomy and Control on the Benefits of Written Emotion Expression," *Personality and Social Psychology Bulletin* 35, no. 3 (2009): 351–64.

77 **these benefits last several months**: Linda D. Cameron and Nickola C. Overall, "Suppression and Expression as Distinct Emotion-Regulation Processes in Daily Interactions: Longitudinal and Meta-Analyses," *Emotion* 18, no. 4 (2018): 465.

77 **Lord Byron once noted**: Lord Byron, *Don Juan*, trans. Benjamin Laroche (BnF Collection, n.p., 2016), e-book.

78 **pain is crucial for our survival**: Donald M. Broom, "Evolution of Pain," *Vlaams Diergeneeskundig Tijdschrift* 70, no. 1 (2001): 17–21; Edgar T. Walters and Amanda C. de Williams, "Evolution of Mechanisms and Behaviour Important for Pain," *Philosophical Transactions of the Royal Society B* 374, no. 1785 (2019): 20190275.

78 **ensure our ancestors' survival or reproduction**: David M. Buss, "The Evolution of Happiness," *American Psychologist* 55, no. 1 (2000): 15.

78 **While physical pain signals**: Broom, "Evolution of Pain," 17–21.

78 **emotional pain indicates**: Notably, both types of pain are processed in the same regions of the brain (Geoff MacDonald and Mark R. Leary, "Why Does Social Exclusion Hurt? The Relationship between Social and Physical Pain," *Psychological Bulletin* 131, no. 2 [2005]: 202).

78 **Sixty-five-year-old Jo Cameron**: Claire Diamond, "The Woman Who Doesn't Feel Pain," BBC Scotland News, March 27, 2019, https://www.bbc.com/news/uk-scotland-highlands-islands-47719718.

78 **"... is part of the solution"**: Randolph M. Nesse and Jay Schulkin, "An Evolutionary Medicine Perspective on Pain and Its Disorders," *Philosophical Transactions of the Royal Society B* 374, no. 1785 (2019): 20190288.

78 **impairs our performance**: June Walker, "Pain and Distraction in Athletes and Non-Athletes," suppl., *Perceptual and Motor Skills* 33, no. 3 (1971): 1187–90.

78 **it's self-deception**: Eccleston and Crombez, "Pain Demands Attention," 356.

78 **Neuroscientists have identified**: Ecole Polytechnique Fédérale de Lausanne, "Brain's Insular Cortex Processes Pain and Drives Learning from Pain," ScienceDaily, May 16, 2019, https://www.sciencedaily.com/releases/2019/05/190516142849.htm.

78 **behavior patterns calcify**: Barbara L. Fredrickson and Marcial F. Losada, "Positive Affect and the Complex Dynamics of Human Flourishing," *American Psychologist* 60, no. 7 (2005): 678–86, at 685.

79 **People who "rise" from challenges**: Brené Brown, *Rising Strong* (New York: Spiegel & Grau, 2017).

83 **our safety system**: Porges, *Pocket Guide to the Polyvagal Theory*.

83 **forgive our bodies**: Slade, "Stephen W. Porges."

84 **they reported greater energy and well-being**: Weinstein and Hodgins, "Moderating Role of Autonomy and Control on the Benefits of Written Emotion Expression," 351–64.

84 **"[our response] becomes our choice"**: Porges, *Pocket Guide to the Polyvagal Theory*, 25.

85 **putting our experiences into words**: University of California–Los Angeles, "Putting Feelings into Words Produces Therapeutic Effects in the Brain," ScienceDaily, June 22, 2007, https://www.sciencedaily.com/releases/2007/06/070622090727.htm.

85 **confronting psychological pain as necessary to healing**: Steven C. Hayes, Kirk D. Strosahl, and Kelly G. Wilson, *Acceptance and Commitment Therapy*, vol. 6 (New York: Guilford Press, 1999).

85 **we start to take away their power**: Steven C. Hayes et al., "Acceptance and Commitment Therapy: Model, Processes and Outcomes," *Behaviour Research and Therapy* 44, no. 1 (2006): 1–25.

Chapter 6: Step 2: Trace Your Triggers

91 **University of Rochester**: Delia O'Hara, "The Intrinsic Motivation of Richard Ryan and Edward Deci," American Psychological Association, 2017, www.apa.org/members/content/intrinsic-motivation.

92 **to shape and optimize it**: Maarten Vansteenkiste and Richard M. Ryan, "On Psychological Growth and Vulnerability: Basic Psychological Need Satisfaction and Need Frustration as a Unifying Principle," *Journal of Psychotherapy Integration* 23, no. 3 (2013): 263.

92 **"apathetic, alienated, and irresponsible"**: Richard M. Ryan and Edward L. Deci, "Self-Determination Theory and the Facilitation of Intrinsic Motivation, Social Development, and Well-Being," *American Psychologist* 55, no. 1 (2000): 68–78.

Notes

92 **"curious, vital, and self-motivated"**: Deci and Ryan go so far as to call the "organismic necessities for health" (Edward L. Deci and Richard M. Ryan, "Motivation, Personality, and Development within Embedded Social Contexts: An Overview of Self-Determination Theory," in *The Oxford Handbook of Human Motivation*, ed. Richard M. Ryan [Oxford: Oxford University Press, 2012], 85–107).

92 **"best" rather than the "beast"**: Vansteenkiste and Ryan, "On Psychological Growth and Vulnerability," 263.

92 **direct path to fulfillment, motivation, growth, and self-actualization**: Richard M. Ryan, Veronika Huta, and Edward L. Deci, "Living Well: A Self-Determination Theory Perspective on Eudaimonia," *Journal of Happiness Studies* 9 (2008): 139–70.

92 **make flourishing virtually impossible**: Ryan and Deci, "Self-Determination Theory and the Facilitation of Intrinsic Motivation, Social Development, and Well-Being," 75.

92 **effective in our actions**: Edward L. Deci and Richard M. Ryan, "Self-Determination Research: Reflections and Future Directions," in *Handbook of Self-Determination Research*, ed. Edward L. Deci and Richard M. Ryan (New York: University of Rochester Press, 2002), 431–41.

92 **capable of achieving our goals**: Frank Martela and Richard M. Ryan, "Clarifying Eudaimonia and Psychological Functioning to Complement Evaluative and Experiential Well-Being: Why Basic Psychological Needs Should Be Measured in National Accounts of Well-Being," *Perspectives on Psychological Science* 18, no. 5 (2023): 17456916221141099.

92 **able to grow and learn new things**: Edward L. Deci and Richard M. Ryan, "The 'What' and 'Why' of Goal Pursuits: Human Needs and the Self-Determination of Behavior," *Psychological Inquiry* 11, no. 4 (2000): 227–68.

92 **The psychological rewards our ancestors got**: Deci and Ryan, "'What' and 'Why' of Goal Pursuits," 227–68.

93 **without pressure or threat**: You Jia Lee, "How Travel Makes Us Happy: The Integrated Travel-Happiness (TH) Model of Cognitive Appraisal Theory and Self-Determination Theory" (PhD diss., University of Guelph, 2018).

93 **with agency and integrity**: Kennon M. Sheldon et al., "Persistent Pursuit of Need-Satisfying Goals Leads to Increased Happiness: A 6-Month Experimental Longitudinal Study," *Motivation and Emotion* 34 (2010): 39–48.

93 **staying true to ourselves**: Marie-Christine Opdenakker, "Need-Supportive and Need-Thwarting Teacher Behavior: Their Importance to Boys' and Girls' Academic Engagement and Procrastination Behavior," *Frontiers in Psychology* 12 (2021): 628064; Pedro Cordeiro et al., "The Portuguese Validation of the Basic Psychological Need Satisfaction and Frustration Scale: Concurrent and Longitudinal Relations to Well-Being and Ill-Being," *Psychologica Belgica* 56, no. 3 (2016): 193.

93 **down perilous paths**: Deci and Ryan, "'What' and 'Why' of Goal Pursuits," 227–68.

93 **forge a life of purpose and meaning**: Martela and Ryan, "Clarifying Eudaimonia and Psychological Functioning to Complement Evaluative and Experiential Well-Being," 17456916221141099.

93 **that we belong**: Deci and Ryan, "Self-Determination Research," 431–41.

93 **get along with others**: E. L. Deci, "Human Needs and the Self-Determination of Behavior Personality," *Psychological Inquiry* 11, no. 4 (2000): 227–68.

93 **mutual closeness and support**: Martela and Ryan, "Clarifying Eudaimonia and Psychological Functioning to Complement Evaluative and Experiential Well-Being," 17456916221141099.

Notes

93 **It's an innate human desire**: Mark R. Leary and R. F. Baumeister, "The Need to Belong," *Psychological Bulletin* 117, no. 3 (1995): 497–529; Deci and Ryan, "'What' and 'Why' of Goal Pursuits," 227–68.

93 **benefit from improved connection**: Mauricio Carvallo and Shira Gabriel, "No Man Is an Island: The Need to Belong and Dismissing Avoidant Attachment Style," *Personality and Social Psychology Bulletin* 32, no. 5 (2006): 697–709; see also Beiwen Chen et al., "Basic Psychological Need Satisfaction, Need Frustration, and Need Strength across Four Cultures," *Motivation and Emotion* 39 (2015): 216–36.

93 **boosts well-being**: Minmin Tang, Dahua Wang, and Alain Guerrien, "A Systematic Review and Meta-Analysis on Basic Psychological Need Satisfaction, Motivation, and Well-Being in Later Life: Contributions of Self-Determination Theory," *PsyCh Journal* 9, no. 1 (2020): 5–33.

93 **life satisfaction**: Tang, Wang, and Guerrien, "Systematic Review and Meta-Analysis on Basic Psychological Need Satisfaction, Motivation, and Well-Being in Later Life," 5–33; Louis Tay and Ed Diener, "Needs and Subjective Well-Being around the World," *Journal of Personality and Social Psychology* 101, no. 2 (2011): 354.

93 **performance**: Christopher P. Cerasoli, Jessica M. Nicklin, and Alexander S. Nassrelgrgawi, "Performance, Incentives, and Needs for Autonomy, Competence, and Relatedness: A Meta-Analysis," *Motivation and Emotion* 40 (2016): 781–813.

93 **stick to healthy habits**: Richard M. Ryan et al., "We Know This Much Is (Meta-Analytically) True: A Meta-Review of Meta-Analytic Findings Evaluating Self-Determination Theory," *Psychological Bulletin* 148, no. 11–12 (2022): 813.

93 **stay engaged**: Sarah-Geneviève Trépanier et al., "On the Psychological and Motivational Processes Linking Job Characteristics to Employee Functioning: Insights from Self-Determination Theory," *Work & Stress* 29, no. 3 (2015): 286–305.

93 **grow through adversity**: Daniel J. Brown, Mustafa Sarkar, and Karen Howells, "Growth, Resilience, and Thriving: A Jangle Fallacy?," in *Growth Following Adversity in Sport: A Mechanism to Positive Change*, ed. Ross Wadey, Melissa Day, and Karen Howells (New York: Routledge, 2020), 59–72.

93 **makes us more self-aware**: Koen Luyckx et al., "Basic Need Satisfaction and Identity Formation: Bridging Self-Determination Theory and Process-Oriented Identity Research," *Journal of Counseling Psychology* 56, no. 2 (2009): 276.

93 **empathetic**: Vansteenkiste and Ryan, "On Psychological Growth and Vulnerability," 263.

93 **cool under pressure**: Richard M. Ryan, Bart Soenens, and Maarten Vansteenkiste, "Reflections on Self-Determination Theory as an Organizing Framework for Personality Psychology: Interfaces, Integrations, Issues, and Unfinished Business," *Journal of Personality* 87, no. 1 (2019): 115–45.

93 **both romantic**: Amy B. Brunell and Gregory D. Webster, "Self-Determination and Sexual Experience in Dating Relationships," *Personality and Social Psychology Bulletin* 39, no. 7 (2013): 970–87; Emilie Eve Gravel, Luc G. Pelletier, and Elke Doris Reissing, "'Doing It' for the Right Reasons: Validation of a Measurement of Intrinsic Motivation, Extrinsic Motivation, and Amotivation for Sexual Relationships," *Personality and Individual Differences* 92 (2016): 164–73.

93 **and platonic**: Richard M. Ryan et al., "Building a Science of Motivated Persons: Self-Determination Theory's Empirical Approach to Human Experience and the Regulation of

Behavior," *Motivation Science* 7, no. 2 (2021): 97; Jennifer G. La Guardia et al., "Within-Person Variation in Security of Attachment: A Self-Determination Theory Perspective on Attachment, Need Fulfillment, and Well-Being," *Journal of Personality and Social Psychology* 79, no. 3 (2000): 367.

93 **enhances classroom experiences**: Vincent F. Filak and Kennon M. Sheldon, "Student Psychological Need Satisfaction and College Teacher-Course Evaluations," *Educational Psychology* 23, no. 3 (2003): 235–47.

93 **reducing conflict and bullying**: Ryan et al., "Building a Science of Motivated Persons," 97.

93 **activating reward centers**: Richard Michael Ryan and Stefano Di Domenico, "Epilogue: Distinct motivations and Their Differentiated Mechanisms: Reflections on the Emerging Neuroscience of Human Motivation," *Recent Developments in Neuroscience Research on Human Motivation* 349 (2016): 349–69. See also Ryan, Soenens, and Vansteenkiste, "Reflections on Self-Determination Theory as an Organizing Framework for Personality Psychology," 115–45.

93 **elevating decision-making**: Stefano I. Di Domenico et al., "In Search of Integrative Processes: Basic Psychological Need Satisfaction Predicts Medial Prefrontal Activation during Decisional Conflict," *Journal of Experimental Psychology: General* 142, no. 3 (2013): 967; Stefano I. Di Domenico et al., "Basic Psychological Needs and Neurophysiological Responsiveness to Decisional Conflict: An Event-Related Potential Study of Integrative Self Processes," *Cognitive, Affective, & Behavioral Neuroscience* 16 (2016): 848–65.

94 **left him emotionally**: Kimberley J. Bartholomew et al., "Self-Determination Theory and Diminished Functioning: The Role of Interpersonal Control and Psychological Need Thwarting," *Personality and Social Psychology Bulletin* 37, no. 11 (2011): 1459–73.

94 **and physically drained**: Bartholomew et al., "Self-Determination Theory and Diminished Functioning," 1459–73; Ryan et al., "Building a Science of Motivated Persons," 97.

94 **burned out, even**: Nicolas Gillet et al., "The Effects of Job Demands and Organizational Resources through Psychological Need Satisfaction and Thwarting," *Spanish Journal of Psychology* 18 (2015): E28; Tiphaine Huyghebaert et al., "Psychological Safety Climate as a Human Resource Development Target: Effects on Workers Functioning through Need Satisfaction and Thwarting," *Advances in Developing Human Resources* 20, no. 2 (2018): 169–81.

94 **more vulnerable to judgmental**: Maarten Vansteenkiste, Christopher P. Niemiec, and Bart Soenens, "The Development of the Five Mini-Theories of Self-Determination Theory: An Historical Overview, Emerging Trends, and Future Directions," in *The Decade Ahead: Theoretical Perspectives on Motivation and Achievement*, ed. Timothy C. Urdan and Stuart A. Karabenick (Bingley, UK: Emerald, 2010), 105–65.

94 **and aggressive behavior**: Ryan et al., "Building a Science of Motivated Persons," 97.

94 **heighten anxiety**: Kristen Harknett, Daniel Schneider, and Rebecca Wolfe, "Losing Sleep over Work Scheduling? The Relationship between Work Schedules and Sleep Quality for Service Sector Workers," *SSM-Population Health* 12 (2020): 100681.

94 **depression**: Bartholomew et al., "Self-Determination Theory and Diminished Functioning," 1459–73; Harknett, Schneider, and Wolfe, "Losing Sleep over Work Scheduling?," 100681.

94 **cynicism**: Majid Murad et al., "The Influence of Despotic Leadership on Counterproductive Work Behavior among Police Personnel: Role of Emotional Exhaustion and Organizational Cynicism," *Journal of Police and Criminal Psychology* 36, no. 3 (2021): 603–15.

Notes

94 **and existential loneliness**: Mehmet Saricali and Deniz Guler, "The Mediating Role of Psychological Need Frustration on the Relationship between Frustration Intolerance and Existential Loneliness," *Current Psychology* 41, no. 8 (2022): 5603–11.

94 **fulfilling our potential**: Nicolas Gillet et al., "The Impact of Organizational Factors on Psychological Needs and Their Relations with Well-Being," *Journal of Business and Psychology* 27 (2012): 437–50.

94 **antisocial behavior**: Ryan et al., "Building a Science of Motivated Persons," 97.

94 **borderline personality disorder**: Jolene Van der Kaap-Deeder, Katrijn Brenning, and Bart Neyrinck, "Emotion Regulation and Borderline Personality Features: The Mediating Role of Basic Psychological Need Frustration," *Personality and Individual Differences* 168 (2021): 110365.

94 **violence and murder**: Richard M. Ryan et al., "Building a Science of Motivated Persons: Self-Determination Theory's Empirical Approach to Human Experience and the Regulation of Behavior," *Motivation Science* 7, no. 2 (2021): 97.

95 **leave us feeling upset**: Steven Poole, "Trigger Warning: How Did 'Triggered' Come to Mean 'Upset,'" *Guardian*, July 25, 2019, https://www.theguardian.com/books/2019/jul/25/trigger-warning-triggered-emotionally-upset-word-of-week.

95 **people with PTSD**: Arlin Cuncic, "What Does It Mean to Be 'Triggered,'" VeryWell Mind, August 23, 2023, https://www.verywellmind.com/what-does-it-mean-to-be-triggered-4175432.

95 **we become hypersensitive**: Sebastiano Costa, Nikos Ntoumanis, and Kimberley J. Bartholomew, "Predicting the Brighter and Darker Sides of Interpersonal Relationships: Does Psychological Need Thwarting Matter?," *Motivation and Emotion* 39 (2015): 11–24.

95 **difficulty controlling our reactions**: Constantin Lagios et al., "Explaining the Negative Consequences of Organizational Dehumanization," *Journal of Personnel Psychology* 21, no. 2 (2021): https://doi.org/10.1027/1866-5888/a000286.

95 **act out of character**: Deci and Ryan, "'What' and 'Why' of Goal Pursuits," 227–68.

95 **crash through avoidance**: Carolyn Spring, "Managing Triggers," *Carolyn Spring* (blog), May 1, 2013, https://www.carolynspring.com/blog/managing-triggers/.

98 **neural circuits that store these memories**: Valérie Doyère et al., "Synapse-Specific Reconsolidation of Distinct Fear Memories in the Lateral Amygdala," *Nature neuroscience* 10, no. 4 (2007): 414–16.

98 **aware of it**: Jacek Debiec, "Memories of Trauma Are Unique Because of How Brains and Bodies Respond to Threat," *The Conversation*, September 24, 2018, https://theconversation.com/memories-of-trauma-are-unique-because-of-how-brains-and-bodies-respond-to-threat-103725.

98 **conditioned threat response**: Mallory E. Bowers and Kerry J. Ressler, "An Overview of Translationally Informed Treatments for Posttraumatic Stress Disorder: Animal Models of Pavlovian Fear Conditioning to Human Clinical Trials," *Biological Psychiatry* 78, no. 5 (2015): E15–E27.

99 **tireless mother, Hetty Wang**: Nathan Chen with Alica Park, *One Jump at a Time: My Story* (New York: Harper, 2022).

99 **"Quad King"**: Jacob Gijy, "Why Is Nathan Chen Called the Quad King? What Does It Mean?," Essentially Sports, February 4, 2022, https://www.essentiallysports.com/beijing-winter-olympics-2022-news-us-sports-news-why-is-nathan-chen-called-the-quad-king-what-does-it-mean.

99 **worst performance of his life**: Christine Brennan, "Figure Skating: Nathan Chen Flops in Olympic Debut with Mistakes in Team Short Program," NorthJersey.com, February 9, 2018, https://www.northjersey.com/story/sports/2018/02/09/figure-skating-nathan-chen-flops-olympic-debut-mistakes-team-short-program/322430002.

100 **"Nathan Chen bombed again"**: Christine Brennan, "Nathan Chen Bombs in His Men's Short Program at 2018 Winter Olympics," USA Today, February 16, 2018, https://www.usatoday.com/story/sports/columnist/brennan/2018/02/16/nathan-chen-bombs-his-mens-short-program-2018-winter-olympics/344126002.

102 **especially sensitive**: Susan M. Bögels and Warren Mansell, "Attention Processes in the Maintenance and Treatment of Social Phobia: Hypervigilance, Avoidance and Self-Focused Attention," *Clinical Psychology Review* 24, no. 7 (2004): 827–56.

102 **blocks or threatens them**: Robert A. Baron, "The Role of Affect in the Entrepreneurial Process," *Academy of Management Review* 33, no. 2 (2008): 328–40.

102 **the more we fixate on it**: Kennon M. Sheldon and Alexander Gunz, "Psychological Needs as Basic Motives, Not Just Experiential Requirements," *Journal of Personality* 77, no. 5 (2009): 1467–92; Henk Aarts, Ap Dijksterhuis, and Peter De Vries, "On the Psychology of Drinking: Being Thirsty and Perceptually Ready," *British Journal of Psychology* 92, no. 4 (2001): 631–42; Fritz Strack and Roland Deutsch, "Reflective and Impulsive Determinants of Social Behavior," *Personality and Social Psychology Review* 8, no. 3 (2004): 220–47.

102 **manipulated or treated unfairly**: Jan-Willem Van Prooijen, "Procedural Justice as Autonomy Regulation," *Journal of Personality and Social Psychology* 96, no. 6 (2009): 1166.

102 **unworthiness**: Jungwoo Ha, "The Impact of Thwarted Competence-Presentation on Turnover Intentions," *Academy of Management Proceedings* 2015, no. 1 (2015): https://doi.org/10.5465/ambpp.2015.14824abstract; Hui Fang et al., "Being Eager to Prove Oneself: U-Shaped Relationship between Competence Frustration and Intrinsic Motivation in Another Activity," *Frontiers in Psychology* 8 (2017): 2123; N. Pontus Leander and Tanya L. Chartrand, "On Thwarted Goals and Displaced Aggression: A Compensatory Competence Model," *Journal of Experimental Social Psychology* 72 (2017): 88–100.

102 **inadequacy**: Kimberley J. Bartholomew et al., "Psychological Need Thwarting in the Sport Context: Assessing the Darker Side of Athletic Experience," *Journal of Sport and Exercise Psychology* 33, no. 1 (2011): 75–102.

102 **inferiority**: Opdenakker, "Need-Supportive and Need-Thwarting Teacher Behavior," 628064; Evangelos Brisimis et al., "Exploring the Relationships of Autonomy-Supportive Climate, Psychological Need Satisfaction and Thwarting with Students' Self-Talk in Physical Education," *Journal of Education, Society and Behavioural Science* 33, no. 11 (2020): 112–22.

103 **Overly demanding tasks**: Joachim Waterschoot, Jolene van der Kaap-Deeder, and Maarten Vansteenkiste, "The Role of Competence-Related Attentional Bias and Resilience in Restoring Thwarted Feelings of Competence," *Motivation and Emotion* 44 (2020): 82–98; Trépanier et al., "On the Psychological and Motivational Processes Linking Job Characteristics to Employee Functioning: Insights from Self-Determination Theory," 286–305.

103 **or expectations**: Kimberley Jane Bartholomew et al., "Job Pressure and Ill-Health in Physical Education Teachers: The Mediating Role of Psychological Need Thwarting," *Teaching and Teacher Education* 37 (2014): 101–7; Chris Giebe and Thomas Rigotti, "Tenets of Self-Determination Theory as a Mechanism behind Challenge Demands: A Within-Person Study," *Journal of Managerial Psychology* 37, no. 5 (2022): 480–97.

103 **time pressure**: Giebe and Rigotti, "Tenets of Self-Determination Theory as a Mechanism behind Challenge Demands," 480–97.
103 **job pressure**: Bartholomew et al., "Job Pressure and Ill-Health in Physical Education Teachers," 101–7.
103 **Having others' approval**: Vello Hein, Andre Koka, and Martin S. Hagger, "Relationships between Perceived Teachers' Controlling Behaviour, Psychological Need Thwarting, Anger and Bullying Behaviour in High-School Students," *Journal of Adolescence* 42 (2015): 103–14.
103 **or attention depending on our performance**: Bartholomew et al., "Self-Determination Theory and Diminished Functioning," 1459–73.
103 **Unchallenging or unrewarding tasks**: Ha, "Impact of Thwarted Competence-Presentation on Turnover Intentions."
103 **Repetitive or boring tasks**: Nicolas Gillet et al., "The Effects of Organizational Factors, Psychological Need Satisfaction and Thwarting, and Affective Commitment on Workers' Well-Being and Turnover Intentions," *Le travail humain* 78, no. 2 (2015): 119–40.
103 **or doing what we're good at**: Ha, "Impact of Thwarted Competence-Presentation on Turnover Intentions."
104 **Unclear or changing standards**: Gillet et al., "Effects of Job Demands and Organizational Resources through Psychological Need Satisfaction and Thwarting," E28.
104 **clarity, certainty, or predictability**: Ryan, Soenens, and Vansteenkiste, "Reflections on Self-Determination Theory as an Organizing Framework for Personality Psychology," 115–45.
104 **Constant changes**: Gillet et al., "Effects of Job Demands and Organizational Resources through Psychological Need Satisfaction and Thwarting," E28.
104 **Unclear expectations or standards**: Gillet et al., "Effects of Organizational Factors, Psychological Need Satisfaction and Thwarting, and Affective Commitment on Workers' Well-Being and Turnover Intentions," 119–40.
104 **Failing**: Waterschoot, van der Kaap-Deeder, and Vansteenkiste, "Role of Competence-Related Attentional Bias and Resilience in Restoring Thwarted Feelings of Competence," 82–98.
104 **"should" have succeeded**: Kennon M. Sheldon and Jonathan C. Hilpert, "The Balanced Measure of Psychological Needs (BMPN) Scale: An Alternative Domain General Measure of Need Satisfaction," *Motivation and Emotion* 36 (2012): 439–51; Evangelos Brisimis et al., "Exploring the Relationships of Autonomy-Supportive Climate, Psychological Need Satisfaction and Thwarting with Students' Self-Talk in Physical Education," *Journal of Education, Society and Behavioural Science* 33, no. 11 (2020): 112–22.
104 **Having our faults highlighted**: Waterschoot, van der Kaap-Deeder, and Vansteenkiste, "Role of Competence-Related Attentional Bias and Resilience in Restoring Thwarted Feelings of Competence," 82–98; Elien Mabbe et al., "The Impact of Feedback Valence and Communication Style on Intrinsic Motivation in Middle Childhood: Experimental Evidence and Generalization across Individual Differences," *Journal of Experimental Child Psychology* 170 (2018): 134–60; Frederick M. E. Grouzet et al., "From Environmental Factors to Outcomes: A Test of an Integrated Motivational Sequence," *Motivation and Emotion* 28 (2004): 331–46.
104 **Being judged, doubted, or criticized**: Ryan, Soenens, and Vansteenkiste, "Reflections on Self-Determination Theory as an Organizing Framework for Personality Psychology," 115–45.
104 **weaknesses pointed out**: Kennon M. Sheldon and Vincent Filak, "Manipulating Autonomy, Competence, and Relatedness Support in a Game-Learning Context: New Evidence That All Three Needs Matter," *British Journal of Social Psychology* 47, no. 2 (2008): 267–83.

Notes

104 **treated as "less than"**: Bartholomew et al., "Psychological Need Thwarting in the Sport Context," 75–102.

104 **acknowledged or appreciated**: Sheldon and Hilpert, "Balanced Measure of Psychological Needs (BMPN) Scale," 439–51.

104 **go against our values, interest, or goals**: Meredith Rocchi et al., "Assessing Need-Supportive and Need-Thwarting Interpersonal Behaviours: The Interpersonal Behaviours Questionnaire (IBQ)," *Personality and Individual Differences* 104 (2017): 423–33; Bartholomew et al., "Psychological Need Thwarting in the Sport Context," 75–102.

104 **Pressure to do what we "should"**: Jochen Delrue et al., "A Game-to-Game Investigation of the Relation between Need-Supportive and Need-Thwarting Coaching and Moral Behavior In Soccer," *Psychology of Sport and Exercise* 31 (2017): 1–10.

104 **Pressure to wear a social mask**: Tiphaine Huyghebaert et al., "Investigating the Longitudinal Effects of Surface Acting on Managers' Functioning through Psychological Needs," *Journal of Occupational Health Psychology* 23, no. 2 (2018): 207.

104 **External factors forcing us to obey**: Tiphaine Huyghebaert et al., "Leveraging Psychosocial Safety Climate to Prevent Ill-Being: The Mediating Role of Psychological Need Thwarting," *Journal of Vocational Behavior* 107 (2018): 111–25.

104 **Being forced**: Rocchi et al., "Assessing Need-Supportive and Need-Thwarting Interpersonal Behaviours," 423–33.

104 **to act against our will**: Sheldon and Hilpert, "Balanced Measure of Psychological Needs (BMPN) Scale," 439–51.

104 **Being prevented from making choices**: Costa, Ntoumanis, and Bartholomew, "Predicting the Brighter and Darker Sides of Interpersonal Relationships," 11–24.

104 **or having our choices limited**: Jochen Delrue et al., "A Game-to-Game Investigation of the Relation between Need-Supportive and Need-Thwarting Coaching and Moral Behavior In Soccer," *Psychology of Sport and Exercise* 31 (2017): 1–10.

104 **Being guilted**: Gillet et al., "Impact of Organizational Factors on Psychological Needs and Their Relations with Well-Being," 437–50.

104 **minimized or invalidated**: Michael H. Kernis, "Toward a Conceptualization of Optimal Self-Esteem," *Psychological Inquiry* 14, no. 1 (2003): 1–26.

104 **Not being listened to or heard or being interrupted**: Tiphaine Huyghebaert-Zouaghi et al., "Advancing the Conceptualization and Measurement of Psychological Need States: A 3 × 3 Model Based on Self-Determination Theory," *Journal of Career Assessment* 29, no. 3 (2021): 396–421.

105 **Unfairness**: C. Daniel Batson et al., "Anger at Unfairness: Is It Moral Outrage?," *European Journal of Social Psychology* 37, no. 6 (2007): 1272–85.

105 **Being excluded from decisions that affect us, having our input or preferences discounted**: Nikita Bhavsar et al., "Conceptualizing and Testing a New Tripartite Measure of Coach Interpersonal Behaviors," *Psychology of Sport and Exercise* 44 (2019): 107–20.

105 **Rejection**: Costa, Ntoumanis, and Bartholomew, "Predicting the Brighter and Darker Sides of Interpersonal Relationships," 11–24.

105 **dismissed**: Bartholomew et al., "Self-Determination Theory and Diminished Functioning," 1459–73.

105 **or abandoned**: Huyghebaert-Zouaghi et al., "Advancing the Conceptualization and Measurement of Psychological Need States," 396–421.

105 **left out**: Opdenakker, "Need-Supportive and Need-Thwarting Teacher Behavior," 628064.

105 **ignored**: Luke Felton and Sophia Jowett, "On Understanding the Role of Need Thwarting in the Association between Athlete Attachment and Well/Ill-Being," *Scandinavian Journal of Medicine & Science in Sports* 25, no. 2 (2015): 289–98; Bhavsar et al., "Conceptualizing and Testing a New Tripartite Measure of Coach Interpersonal Behaviors," 107–20.

105 **Being actively avoided**: Bhavsar et al., "Conceptualizing and Testing a New Tripartite Measure of Coach Interpersonal Behaviors," 107–20.

105 **or disliked**: Huyghebaert-Zouaghi et al., "Advancing the Conceptualization and Measurement of Psychological Need States," 396–421.

105 **Connections**: David M. Buss, "The Evolution of Happiness," *American Psychologist* 55, no. 1 (2000): 15.

105 **Lack of care or concern**: Rocchi et al., "Assessing Need-Supportive and Need-Thwarting Interpersonal Behaviours," 423–33; Filipe Rodrigues et al., "The Role of Dark-Side of Motivation and Intention to Continue in Exercise: A Self-Determination Theory Approach," *Scandinavian Journal of Psychology* 60, no. 6 (2019): 585–95.

105 **Being treated coldly**: Ryan, Soenens, and Vansteenkiste, "Reflections on Self-Determination Theory as an Organizing Framework for Personality Psychology," 115–45.

105 **inattentively**: Costa, Ntoumanis, and Bartholomew, "Predicting the Brighter and Darker Sides of Interpersonal Relationships," 11–24; Bartholomew et al., "Self-Determination Theory and Diminished Functioning," 1459–73.

105 **indifferently**: Ryan, Soenens, and Vansteenkiste, "Reflections on Self-Determination Theory as an Organizing Framework for Personality Psychology," 115–45.

105 **Lack of support**: Rocchi et al., "Assessing Need-Supportive and Need-Thwarting Interpersonal Behaviours," 423–33.

105 **Concern**: Gillet et al., "Impact of Organizational Factors on Psychological Needs and Their Relations with Well-Being," 437–50.

105 **Others being favored over us**: Buss, "Evolution of Happiness," 15.

105 **Tension with important others**: Sheldon and Hilpert, "Balanced Measure of Psychological Needs (BMPN) Scale," 439–51.

105 **Dehumanizing or hurtful treatment**: Lei Cheng et al., "Objectification Limits Authenticity: Exploring the Relations between Objectification, Perceived Authenticity, and Subjective Well-Being," *British Journal of Social Psychology* 61, no. 2 (2022): 622–43.

105 **Being punished, blamed**: Tiphaine Huyghebaert-Zouaghi et al., "Managerial Predictors and Motivational Outcomes of Workers' Psychological Need States Profiles: A Two-Wave Examination," *European Journal of Work and Organizational Psychology* 32, no. 2 (2023): 216–33; Nikita Bhavsar et al., "Conceptualizing and Testing a New Tripartite Measure of Coach Interpersonal Behaviors," *Psychology of Sport and Exercise* 44 (2019): 107–20.

105 **humiliated or sabotaged**: Hein, Koka, and Hagger, "Relationships between Perceived Teachers' Controlling Behaviour, Psychological Need Thwarting, Anger and Bullying Behaviour in high-School Students," 103–14.

105 **Abuse, microaggressions, or passive aggressiveness**: Hein, Koka, and Hagger, "Relationships between Perceived Teachers' Controlling Behaviour, Psychological Need Thwarting, Anger and Bullying Behaviour in high-School Students," 103–14.

105 **Feeling objectified or used**: Cheng et al., "Objectification Limits Authenticity," 622–43; Murad et al., "Influence of Despotic Leadership on Counterproductive Work Behavior among Police Personnel: Role of Emotional Exhaustion and Organizational Cynicism," *Journal of Police and Criminal Psychology* 36, no. 3 (2021): 603–15; Hein, Koka, and Hagger,

"Relationships between Perceived Teachers' Controlling Behaviour, Psychological Need Thwarting, Anger and Bullying Behaviour in high-School Students," 103–14.

105 **Betrayal**: Buss, "Evolution of Happiness," 15.

CHAPTER 7: STEP 3: SPOT YOUR SHADOWS

106 **family lineage boasting philosophers**: Deirdre Bair, *Jung: A Biography* (New York: Back Bay Books, 2004).

106 **set him apart from his peers**: Michael S. M. Fordham and Frieda Fordham, "Character of His Psychotherapy," in "Carl Jung" (main entry), *Ecyclopedia Britannica*, last updated July 22, 2024, https://www.britannica.com/biography/Carl-Jung/Character-of-his-psychotherapy.

106 **his vivid imagination**: Bair, *Jung*.

106 **profession of psychiatry**: Wayne Viney and D. Brett King, *A History of Psychology: Ideas and Context*, 3rd ed. (Boston, MA: Allyn and Bacon, 2003).

107 **with his betrothal to Emma Rauschenbach**: Lucy Scholes, "Review: Labyrinths: Emma Jung, Her Marriage to Carl, and the Early Years of Psychoanalysis by Catrina Clay," *Guardian*, August 7, 2016, https://www.theguardian.com/books/2016/aug/07/labyrinths-emma-jung-marriage-carl-early-years-psychoanalysis-catrine-clay-review.

107 **the (by then) famous founder of psychotherapy**: Mark D. Kelland, "A Brief Biography of Carl Jung," chap. 13.2 of *Personality Theory in a Cultural Context*, LibreTexts: Social Sciences, accessed September 11, 2024, https://socialsci.libretexts.org/Bookshelves/Psychology/Culture_and_Community/Personality_Theory_in_a_Cultural_Context_(Kelland)/13%3A_Carl_Jung/13.02%3A_A_Brief_Biography_of_Carl_Jung.

107 **collaboration bound for combustion**: Bair, *Jung*.

107 **Freud's desire for an intellectual heir**: Society of Analytical Psychology, "About Carl Jung," accessed September 11, 2024, https://www.thesap.org.uk/articles-on-jungian-psychology-2/carl-gustav-jung.

107 **ill-suited for discipleship**: Bair, *Jung*.

107 **tame his own neuroses**: Chris Allen, "Carl Jung," chap. 5 of *The Balance of Personality*, open textbook, Portland State University, https://pdx.pressbooks.pub/thebalanceofpersonality/chapter/chapter-5-carl-jung.

107 **lingering effect of past disappointments**: Bair, *Jung*, 238.

107 **"failed bromance"**: Lloyd I. Sederer, MD, "The Failed 'Bromance' between Sigmund Freud and Carl Jung," Psychology Today, December 5, 2021, https://www.psychologytoday.com/us/blog/therapy-it-s-more-just-talk/202112/the-failed-bromance-between-sigmund-freud-and-carl-jung.

107 **scholar into a profound crisis**: Bair, *Jung*.

107 **aptly named "black books"**: Victor Bodo, "Beyond Freud: Carl Jung's Lasting Influence on Psychology," Medium, September 19, 2023, https://medium.com/@dr.victor.bodo/beyond-freud-carl-jungs-lasting-influence-on-psychology-7d7264f78907.

107 **respected empirical scientist**: Bair, *Jung*.

108 **shape our ideas, emotions, and actions**: It covers various aspects of the self like the ego, persona, animus, and personal and collective unconscious (Carl Gustav Jung, *Two Essays on Analytical Psychology* [New York: Routledge, 2014]).

108 **giving in to our "shadows"**: C. G. Jung, *Collected Works of C. G. Jung*, vol. 11, *Psychology and Religion: West and East*, ed. Gerhard Adler and R. F. C. Hull (Princeton, NJ: Princeton

University Press, 1969), 76; vol. 16, *Practice of Psychotherapy* (Princeton, NJ: Princeton University Press, 2014), paragraph 470.

109 **"quick fixes"**: Nele Laporte et al., "Adolescents as Active Managers of Their Own Psychological Needs: The Role of Psychological Need Crafting in Adolescents' Mental Health," *Journal of Adolescence* 88 (2021): 67–83.

109 **self-worth**: Crocker, Jennifer, and Connie T. Wolfe, "Contingencies of Self-Worth," *Psychological Review* 108.3 (2001): 593.

109 **power, or approval**: Edward L. Deci and Richard M. Ryan, "The 'What' and 'Why' of Goal Pursuits: Human Needs and the Self-Determination of Behavior," *Psychological Inquiry* 11, no. 4 (2000): 227–68.

109 **not all goals are created equal**: Maarten Vansteenkiste, Christopher P. Niemiec, and Bart Soenens, "The Development of the Five Mini-Theories of Self-Determination Theory: An Historical Overview, Emerging Trends, and Future Directions," in *The Decade Ahead: Theoretical Perspectives on Motivation and Achievement*, ed. Timothy C. Urdan and Stuart A. Karabenick (Bingley, UK: Emerald, 2010), 105–65.

109 **In this analogy**: Deci and Ryan. "The 'What' and 'Why' of Goal Pursuits," 227–26; Frederick L. Philippe et al., "Work-Related Episodic Memories Can Increase or Decrease Motivation and Psychological Health at Work," *Work & Stress* 33, no. 4 (2019): 366–84.

109 **frustration, anxiety, and insecurity that unmet needs create**: Maarten Vansteenkiste and Richard M. Ryan, "On Psychological Growth and Vulnerability: Basic Psychological Need Satisfaction and Need Frustration as a Unifying Principle," *Journal of Psychotherapy Integration* 23, no. 3 (2013): 263.

109 **as tempting as carbs**: Vansteenkiste, Niemiec, and Soenens, "Development of the Five Mini-Theories of Self-Determination Theory," 105–65.

109 **But no matter how much shadow goals**: Deci and Ryan, "'What' and 'Why' of Goal Pursuits," 227–68.

109 **spin our wheels**: Vansteenkiste, Niemiec, and Soenens, "Development of the Five Mini-Theories of Self-Determination Theory," 105–65.

109 **what we really need to thrive**: Richard M. Ryan and Edward L. Deci, "A Self-Determination Theory Approach to Psychotherapy: The Motivational Basis for Effective Change," *Canadian Psychology/Psychologie canadienne* 49, no. 3 (2008): 186.

109 **three compensatory motives**: Deci and Ryan, "'What' and 'Why' of Goal Pursuits," 227–68.

110 **we might become defensive**: Kristin D. Neff, Ya-Ping Hsieh, and Kullaya Dejitterat, "Self-Compassion, Achievement Goals, and Coping with Academic Failure," *Self and Identity* 4, no. 3 (2005): 263–87.

110 **try to reestablish control**: Maarten Vansteenkiste et al., "Examining Correlates of Game-to-Game Variation in Volleyball Players' Achievement Goal Pursuit and Underlying Autonomous and Controlling Reasons," *Journal of Sport and Exercise Psychology* 36, no. 2 (2014): 131–45; James N. Donald et al., "Paths to the Light and Dark Sides of Human Nature: A Meta-Analytic Review of the Prosocial Benefits of Autonomy and the Antisocial Costs of Control," *Psychological Bulletin* 147, no. 9 (2021): 921.

110 **punish those who've hurt us**: Roy F. Baumeister et al., "Thwarting the Need to Belong: Understanding the Interpersonal and Inner Effects of Social Exclusion," *Social and Personality Psychology Compass* 1, no. 1 (2007): 506–20.

Notes

110 **they externalize our problems**: Joke Verstuyf et al., "Motivational Dynamics of Eating Regulation: A Self-Determination Theory Perspective," *International Journal of Behavioral Nutrition and Physical Activity* 9, no. 1 (2012): 1–16.

110 **the motive of proving**: Vansteenkiste and Ryan, "On Psychological Growth and Vulnerability," 263.

110 **despite our fears**: Deci and Ryan, "'What' and 'Why' of Goal Pursuits," 227–68; see also Trish Gorely et al., "Nutrition and Physical Activity," *International Journal of Behavioral Nutrition and Physical Activity* 6 (2009): 5.

110 **works hard to achieve**: Deci and Ryan, "'What' and 'Why' of Goal Pursuits," 227–68; Kennon M. Sheldon and Tim Kasser, "Psychological Threat and Extrinsic Goal Striving," *Motivation and Emotion* 32 (2008): 37–45; Tim Kasser and Richard M. Ryan, "A Dark Side of the American Dream: Correlates of Financial Success as a Central Life Aspiration, *Journal of Personality and Social Psychology* 65, no. 2 (1993): 410.

110 **or seek external rewards**: Tim Kasser and Richard M. Ryan, "Further Examining the American Dream: Differential Correlates of Intrinsic and Extrinsic Goals," *Personality and Social Psychology Bulletin* 22, no. 3 (1996): 280–87.

110 **to gain popularity or status**: Murray, *Explorations in Personality*.

110 **final compensatory motive, preventing**: Susan M. Bögels and Warren Mansell, "Attention Processes in the Maintenance and Treatment of Social Phobia: Hypervigilance, Avoidance and Self-Focused Attention," *Clinical Psychology Review* 24, no. 7 (2004): 827–56.

110 **stop our momentary mood**: Deci and Ryan, "'What' and 'Why' of Goal Pursuits," 227–68.

110 **escaping**: Todd F. Heatherton and Roy F. Baumeister, "Binge Eating as Escape from Self-Awareness," *Psychological Bulletin* 110, no. 1 (1991): 86.

110 **ignoring**: Bögels and Mansell, "Attention Processes in the Maintenance and Treatment of Social Phobia," 827–56; Rémi Radel et al., "The Role of (Dis)Inhibition in Creativity: Decreased Inhibition Improves Idea Generation," *Cognition* 134 (2015): 110–20; Laporte et al., "Adolescents as Active Managers of Their Own Psychological Needs," 67–83.

110 **or downplaying issues**: Deci and Ryan, "'What' and 'Why' of Goal Pursuits," 227–68.

110 **opt out of situations**: Henry A. Murray, *Explorations in Personality* (New York: Oxford University Press, 1938).

110 **giving up or giving in**: Deci and Ryan, "'What' and 'Why' of Goal Pursuits," 227–68; Kimberley J. Bartholomew et al., "Psychological Need Thwarting in the Sport Context: Assessing the Darker Side of Athletic Experience," *Journal of Sport and Exercise Psychology* 33, no. 1 (2011): 75–102; Vansteenkiste and Ryan, "On Psychological Growth and Vulnerability," 263.

110 **distance ourselves from others**: Murray, *Explorations in Personality*.

110 **short-term escape**: Verstuyf et al., "Motivational Dynamics of Eating Regulation," 1–16.

111 **a sense of authentic choice**: Kasser and Ryan, "Further Examining the American Dream," 280–87; C. Raymond Knee et al., "Self-Determination as Growth Motivation in Romantic Relationships," *Personality and Social Psychology Bulletin* 28, no. 5 (2002): 609–19.

111 **out of pressure**: Richard M. Ryan, Bart Soenens, and Maarten Vansteenkiste, "Reflections on Self-Determination Theory as an Organizing Framework for Personality Psychology: Interfaces, Integrations, Issues, and Unfinished Business," *Journal of Personality* 87, no. 1 (2019): 115–45.

Notes

111 **expectation for rewards**: Vansteenkiste, Niemiec, and Soenens, "Development of the Five Mini-Theories of Self-Determination Theory," 105–65.

111 **And while the latter**: This mini-theory of SDT is called cognitive evaluation theory. For a review, see Vansteenkiste, Niemiec, and Soenens, "Development of the Five Mini-Theories of Self-Determination Theory," 105–65.

111 **actively distances us from them**: Christopher P. Niemiec, Richard M. Ryan, and Edward L. Deci, "The Path Taken: Consequences of Attaining Intrinsic and Extrinsic Aspirations in Post-College Life," *Journal of Research in Personality* 43, no. 3 (2009): 291–306; see also Deci and Ryan, "'What' and 'Why' of Goal Pursuits," 227–68; Richard M. Ryan et al., "All Goals Are Not Created Equal: An Organismic Perspective on the Nature of Goals and Their Regulation," in *The Psychology of Action: Linking Cognition and Motivation to Behavior*, ed. P. M. Gollwitzer and J. A. Bargh (New York: Guilford Press, 1996); Richard M. Ryan, Veronika Huta, and Edward L. Deci, "Living Well: A Self-Determination Theory Perspective on Eudaimonia," *Journal of Happiness Studies* 9 (2008): 139–70.

111 **Shadow goals are overwhelmingly extrinsic**: Rémi Radel et al., "Restoration Process of the Need for Autonomy: the Early Alarm Stage," *Journal of Personality and Social Psychology* 101, no. 5 (2011): 919.

111 **downplay**: Deci and Ryan, "'What' and 'Why' of Goal Pursuits," 227–68.

111 **compromise**: Paulo N. Vieira et al., "Predictors of Psychological Well-Being during Behavioral Obesity Treatment in Women," *Journal of Obesity* (2011).

111 **with lesser alternatives**: Laporte et al., "Adolescents as Active Managers of Their Own Psychological Needs," 67–83; Ryan and Deci, "Self-Determination Theory Approach to Psychotherapy," 186.

111 **cycle of pressure, negativity, and ill-being**: Michael H. Kernis, "Author's Response: Optimal Self-Esteem and Authenticity: Separating Fantasy from Reality," *Psychological Inquiry* 14, no. 1 (2003): 83–89; Deci and Ryan, "'What' and 'Why' of Goal Pursuits," 227–68; Tim Kasser et al., "The Relations of Maternal and Social Environments to Late Adolescents' Materialistic and Prosocial Values," *Developmental Psychology* 31, no. 6 (1995): 907.

111 **rewire our brains**: Ryan, Soenens, and Vansteenkiste, "Reflections on Self-Determination Theory as an Organizing Framework for Personality Psychology," 115–45.

111 **poor replacements**: Vansteenkiste, Niemiec, and Soenens, "Development of the Five Mini-Theories of Self-Determination Theory," 105–65.

111 **dubbing him "America's hope"**: Jim Caple, "How U.S. Teen Nathan Chen Jumped into Upper Echelon of Figure Skating," ESPN, March 29, 2017, https://www.espn.com/olympics/figureskating/story/_/id/19017406/how-teen-figure-skater-nathan-chen-us-national-champion-spun-quad-jumps-gold.

111 **a "new kind of pressure"**: Nathan Chen with Alica Park, *One Jump at a Time: My Story* (New York: Harper, 2022), 80.

111 **powerfully undermining his confidence**: Vansteenkiste, Niemiec, and Soenens, "Development of the Five Mini-Theories of Self-Determination Theory," 105–65.

111 **But with each passing day**: Chen with Park. *One Jump at a Time*, 216.

112 **proving his competence**: Vansteenkiste, Niemiec, and Soenens, "Development of the Five Mini-Theories of Self-Determination Theory," 105–65.

112 **what was I worth**: Chen with Park. *One Jump at a Time*, 90; Jennifer Crocker and Connie T. Wolfe, "Contingencies of Self-Worth," *Psychological Review* 108, no. 3 (2001): 593.

Notes

112 **die for an Olympic gold medal**: Juan Marcos González et al., "Trading Health Risks for Glory: A Reformulation of the Goldman dilemma," *Sports Medicine* 48 (2018): 1963–69.

113 **that didn't depend on winning**: Kernis, "Author's Response," 83–89.

113 **becoming the first American figure skater**: Scott Stump, "Why US Figure Skaters from 2022 Winter Olympics Are Just Now Getting Their Medals," *Today*, January 31, 2024, https://www.today.com/news/sports/us-figure-skaters-gold-medals-2022-winter-olympics-rcna136528.

114 **how shadow goals want us to act**: Deci and Ryan, "'What' and 'Why' of Goal Pursuits," 227–68.

115 **making the space for self-acceptance and reinvention**: Christopher Perry, "The Jungian Shadow," About Jungian Analysis and Psychotherapy, Society of Analytical Psychology, accessed September 11, 2024, https://www.thesap.org.uk/articles-on-jungian-psychology-2/about-analysis-and-therapy/the-shadow.

117 **A drive to achieve greatness**: Kennon M. Sheldon and Tim Kasser, "Psychological Threat and Extrinsic Goal Striving," *Motivation and Emotion* 32 (2008): 37–45; Kasser and Ryan, "Dark Side of the American Dream," 410; Vansteenkiste, Niemiec, and Soenens, "Development of the Five Mini-Theories of Self-Determination Theory," 105–65.

117 **A drive to amass material rewards**: Ryan, Huta, and Deci, "Living Well," 139–70; Kasser and Ryan, "Further Examining the American Dream," 280–87.

117 **A drive to meet excessively high standards**: Sarah H. Mallinson and Andrew P. Hill, "The Relationship between Multidimensional Perfectionism and Psychological Need Thwarting in Junior Sports Participants," *Psychology of Sport and Exercise* 12, no. 6 (2011): 676–84.

117 **A drive to exercise control**: Added to aspirations index by Richard M. Ryan et al., "The American Dream in Russia: Extrinsic Aspirations and Well-Being in Two Cultures," *Personality and Social Psychology Bulletin* 25, no. 12 (1999): 1509–24; see also Ryan, Huta, and Deci, "Living Well," 139–70.

117 **A drive to control or limit**: Deci and Ryan, "'What' and 'Why' of Goal Pursuits," 227–68; Verstuyf et al., "Motivational Dynamics of Eating Regulation," 1–16.

117 **A drive to be liked**: Ryan et al., "American Dream in Russia," 1509–24; Allison M. Ryan and Sungok Serena Shim, "Social Achievement Goals: The Nature and Consequences of Different Orientations toward Social Competence," *Personality and Social Psychology Bulletin* 32, no. 9 (2006): 1246–63.

117 **A drive to be accepted by others**: Alain Van Hiel and Maarten Vansteenkiste, "Ambitions Fulfilled? The Effects of Intrinsic and Extrinsic Goal Attainment on Older Adults' Ego-Integrity and Death Attitudes," *International Journal of Aging and Human Development* 68, no. 1 (2009): 27–51.

117 **A drive to blindly defy**: Vansteenkiste and Ryan, "On Psychological Growth and Vulnerability," 263; Maarten Vansteenkiste et al., "Examining Correlates of Game-to-Game Variation in Volleyball Players' Achievement Goal Pursuit and Underlying Autonomous and Controlling Reasons," 131–45; Earl et al., "Autonomy and Competence Frustration in Young Adolescent Classrooms," 32–40; Donald et al., "Paths to the Light and Dark Sides of Human Nature," 921.

117 **A drive to hurt others**: Yaniv Kanat-Maymon et al., "The Role of Basic Need Fulfillment in Academic Dishonesty: A Self-Determination Theory Perspective," *Contemporary Educational Psychology* 43 (2015): 1–9; Julien S. Bureau et al., "Investigating How Autonomy-Supportive Teaching Moderates the Relation between Student Honesty and Premeditated Cheating,"

British Journal of Educational Psychology 92, no. 1 (2022): 175–93; Murray, *Explorations in Personality*; Richard M. Ryan et al., "Building a Science of Motivated Persons: Self-Determination Theory's Empirical Approach to Human Experience and the Regulation of Behavior," *Motivation Science* 7, no. 2 (2021): 97.

117 **A drive to ignore**: Laporte et al., "Adolescents as Active Managers of Their Own Psychological Needs," 67–83; Verstuyf et al., "Motivational Dynamics of Eating Regulation," 1–16.

117 **A drive to do what we "ought"**: Laporte et al., "Adolescents as Active Managers of Their Own Psychological Needs," 67–83.

117 **A drive to surrender**: Vansteenkiste and Ryan, "On Psychological Growth and Vulnerability," 263; Deci and Ryan, "'What' and 'Why' of Goal Pursuits," 227–68.

117 **A drive to avoid rocking the boat**: Daniel Freeman et al., "Psychological Investigation of the Structure of Paranoia in a Non-Clinical Population," *British Journal of Psychiatry* 186, no. 5 (2005): 427–35.

117 **A drive to close ourselves off**: Kennon M. Sheldon and Alexander Gunz, "Psychological Needs as Basic Motives, Not Just Experiential Requirements," *Journal of Personality* 77, no. 5 (2009): 1467–92.

CHAPTER 8: STEP 4: PICK YOUR PIVOTS

119 **proactively moving away**: Sharon K. Parke, Uta K. Bindl, and Karoline Strauss, "Making Things Happen: A Model of Proactive Motivation," *Journal of Management* 36, no. 4 (2010): 827–56; Nele Laporte et al., "Testing an Online Program to Foster Need Crafting during the COVID-19 Pandemic," *Current Psychology*, March 28, 2022, 1–18.

119 **the need to immediately investigate and course correct**: The Joint Commission, "Sentinel Event Policy and Procedures," accessed September 11, 2024, https://www.jointcommission.org/resources/sentinel-event/sentinel-event-policy-and-procedures.

121 **parenting styles, teacher behavior**: Richard M. Ryan et al., "We Know This Much Is (Meta-Analytically) True: A Meta-Review of Meta-Analytic Findings Evaluating Self-Determination Theory," *Psychological Bulletin* 148, no. 11–12 (2022): 813; see also Vello Hein, Andre Koka, and Martin S. Hagger, "Relationships between Perceived Teachers' Controlling Behaviour, Psychological Need Thwarting, Anger and Bullying Behaviour in High-School Students," *Journal of Adolescence* 42 (2015): 103–14.

121 **and workplace dynamics**: Nicolas Gillet et al., "The Impact of Organizational Factors on Psychological Needs and Their Relations with Well-Being," *Journal of Business and Psychology* 27 (2012): 437–50; Nicolas Gillet et al., "The Effects of Job Demands and Organizational Resources through Psychological Need Satisfaction and Thwarting," *Spanish Journal of Psychology* 18 (2015): E28; Sarah-Geneviève Trépanier et al., "On the Psychological and Motivational Processes Linking Job Characteristics to Employee Functioning: Insights from Self-Determination Theory," *Work & Stress* 29, no. 3 (2015): 286–305.

121 **we possess the power**: Nele Laporte et al., "Say Hi to Need Crafting: The Pro-Active Side of Need Based Functioning in Adolescence: The Introduction of Need Crafting in a Cross-Sectional and Longitudinal Study," Self-Determination Conference, Egmond aan Zee, May 21–24, 2019.

121 **choosing new goals and habits**: Laporte et al., "Testing an Online Program to Foster Need Crafting during the COVID-19 Pandemic," 1–18.

121 **Laporte's study of over eight hundred adolescents**: Laporte et al., "Say Hi to Need Crafting."

Notes

121 **but also during challenge and crisis**: Netta Weinstein, Farah Khabbaz, and Nicole Legate, "Enhancing Need Satisfaction to Reduce Psychological Distress in Syrian Refugees," *Journal of Consulting and Clinical Psychology* 84, no. 7 (2016): 645.

121 **a study during COVID lockdowns**: Behzad Behzadnia and Saeideh FatahModares, "Basic Psychological Need-Satisfying Activities during the COVID-19 Outbreak," *Applied Psychology: Health and Well-Being* 12, no. 4 (2020): 1115–13.

121 **like financial hardship**: Louis Tay and Ed Diener, "Needs and Subjective Well-Being around the World," *Journal of Personality and Social Psychology* 101, no. 2 (2011): 354.

121 **dangerous living conditions**: Beiwen Chen et al., "Does Psychological Need Satisfaction Matter When Environmental or Financial Safety Are at Risk?," *Journal of Happiness Studies* 16, no. 3 (2015): 745–66.

121 **displacement due to war**: Weinstein, Khabbaz, and Legate, "Enhancing Need Satisfaction to Reduce Psychological Distress in Syrian Refugees," 645.

121 **We have the power**: Sharon K. Parker, Uta K. Bindl, and Karoline Strauss, "Making Things Happen: A Model of Proactive Motivation," *Journal of Management* 36, no. 4 (2010): 827–56; Laporte et al., "Testing an Online Program to Foster Need Crafting during the COVID-19 Pandemic," 1–18.

122 **across multiple samples and situations**: By "multiple situations," I am referring to one study that was a between-subjects design where participants listed three "bad things" and their responses to each.

122 **you can focus on boosting your confidence**: Weinstein, Khabbaz, and Legate, "Enhancing Need Satisfaction to Reduce Psychological Distress in Syrian Refugees," 645.

122 **hobby outside work**: Weinstein, Khabbaz, and Legate, "Enhancing Need Satisfaction to Reduce Psychological Distress in Syrian Refugees," 645.

122 **reinvest in another one**: Kipling D. Williams, "Ostracism: A Temporal Need-Threat Model," *Advances in Experimental Social Psychology* 41 (2009): 275–314.

124 **many "right ways"**: Kennon M. Sheldon et al., "Persistent Pursuit of Need-Satisfying Goals Leads to Increased Happiness: A 6-month Experimental Longitudinal Study," *Motivation and Emotion* 34 (2010): 39–48.

124 **that are personally important**: Behzad Behzadnia and Saeideh FatahModares, "Basic Psychological Need-Satisfying Activities during the COVID-19 Outbreak," *Applied Psychology: Health and Well-Being* 12, no. 4 (2020): 1115–39.

124 **interesting**: Maarten Vansteenkiste, Christopher P. Niemiec, and Bart Soenens, "The Development of the Five Mini-Theories of Self-Determination Theory: An Historical Overview, Emerging Trends, and Future Directions," in *The Decade Ahead: Theoretical Perspectives on Motivation and Achievement*, ed. Timothy C. Urdan and Stuart A. Karabenick (Bingley, UK: Emerald, 2010), 105–65.

124 **enjoyable**: Kennon M. Sheldon and Tim Kasser, "Coherence and Congruence: Two Aspects of Personality Integration," *Journal of Personality and Social Psychology* 68, no. 3 (1995): 531; Richard M. Ryan et al., "All Goals Are Not Created Equal: An Organismic Perspective on the Nature of Goals and Their Regulation," in *The Psychology of Action: Linking Cognition and Motivation to Behavior*, ed. P. M. Gollwitzer and J. A. Bargh (New York: Guilford Press, 1996); Maarten Vansteenkiste et al., "Motivating Learning, Performance, and Persistence: The Synergistic Effects of Intrinsic Goal Contents and Autonomy-Supportive Contexts," *Journal of Personality and Social Psychology* 87, no. 2 (2004): 246.

124 **challenging (in a good way)**: Christopher P. Cerasoli and Michael T. Ford, "Intrinsic Motivation, Performance, and the Mediating Role of Mastery Goal Orientation: A Test of Self-Determination Theory," *Journal of Psychology* 148, no. 3 (2014): 267–86.

124 **trying to change**: Richard M. Ryan and Edward L. Deci, "Intrinsic and Extrinsic Motivations: Classic Definitions and New Directions," *Contemporary Educational Psychology* 25, no. 1 (2000): 54–67; Maarten Vansteenkiste et al., "Motivational Dynamics among Eating-Disordered Patients with and without Nonsuicidal Self-Injury: A Self-Determination Theory Approach," *European Eating Disorders Review* 21, no. 3 (2013): 209–14; Miron Zuckerman et al., "On the Importance of Self-Determination for Intrinsically-Motivated Behavior," *Personality and Social Psychology Bulletin* 4, no. 3 (1978): 443–46.

124 **over a six-month period**: Kennon M. Sheldon et al., "Persistent Pursuit of Need-Satisfying Goals Leads to Increased Happiness: A 6-Month Experimental Longitudinal Study," *Motivation and Emotion* 34 (2010): 39–48.

127 **"hitting the literature"**: This is a great quote from Dr. Lawrence Afrin, one of the leading voices in treating one of my diseases, mast cell activation syndrome (Lawrence B. Afrin, *Never Bet against Occam: Mast Cell Activation Disease and the Modern Epidemics of Chronic Illness and Medical Complexity* [Bethesda, MD: Sisters Media, 2016]).

130 **ten to twenty horrifying years**: Laura Kiesel, "Ehlers-Danlos Syndrome: A Mystery Solved," *Harvard Health Blog*, August 7, 2017, https://www.health.harvard.edu/blog/ehlers-danlos-syndrome-mystery-solved-2017080712122.

130 **listed the telltale signs**: Anne Martin, "An Acquired or Heritable Connective Tissue Disorder? A Review of Hypermobile Ehlers Danlos Syndrome," *European Journal of Medical Genetics* 62, no. 7 (2019): 103672.

133 **scientifically supported actions**: Weinstein, Khabbaz, and Legate, "Enhancing Need Satisfaction to Reduce Psychological Distress in Syrian Refugees," 645.

Chapter 9: Crafting Confidence

137 **$69 million budget**: Ben Fritz, "Movie Projector: 'Schmucks,' Cats, Dogs and Zac Efron Will All Open behind 'Inception,'" *Company Town Blog* (*Los Angeles Times*), July 29, 2010, https://www.latimes.com/archives/blogs/company-town-blog/story/2010-07-29/movie-projector-schmucks-cats-dogs-and-zac-efron-will-all-open-behind-inception.

138 **confidence has two elements**: Shi Yu et al., "Doing Well vs. Doing Better: Preliminary Evidence for the Differentiation of the 'Static' and 'Incremental' Aspects of the Need for Confidence," *Journal of Happiness Studies* 23, no. 3 (2022): 1121–41.

138 **unrelated to our actual abilities**: Edward L. Deci and Richard M. Ryan, "Self-Determination Research: Reflections and Future Directions," in *Handbook of Self-Determination Research*, ed. Edward L. Deci and Richard M. Ryan (New York: University of Rochester Press, 2002).

138 **Get your scrawny ass out of here**: Joanne Rosa, "Josh Groban Opens Up about Anxiety, Depression Early in Career: 'I Was Terrified' and 'Full of Self-Doubt,'" ABC News, December 5, 2019, https://abcnews.go.com/Entertainment/josh-groban-opens-anxiety-depression-early-career-terrified/story?id=6751853.

139 **athletic ability**: An De Meester et al., "Identifying Profiles of Actual and Perceived Motor Confidence among Adolescents: Associations with Motivation, Physical Activity, and Sports Participation," *Journal of Sports Sciences* 34, no. 21 (2016): 2027–37; Laura Bortoli et al.,

Notes

"Confidence, Achievement Goals, Motivational Climate, and Pleasant Psychobiosocial States in Youth Sport," *Journal of Sports Sciences* 29, no. 2 (2011): 171–80.

139 **academic performance**: Shi Yu et al., "Self-Determined Motivation to Choose College Majors, Its Antecedents, and Outcomes: A Cross-Cultural Investigation," *Journal of Vocational Behavior* 108 (2018): 132–50.

139 **financial know-how**: Kate Mielitz, "Using Self-Determination Theory to Investigate Financial Well-Being," *Consumer Interests Annual* 67 (2021).

139 **most reliably predicts our well-being**: Mielitz, "Using Self-Determination Theory to Investigate Financial Well-Being."

139 **confidence can take a nosedive**: Tiphaine Huyghebaert-Zouaghi et al., "Managerial Predictors and Motivational Outcomes of Workers' Psychological Need States Profiles: A Two-Wave Examination," *European Journal of Work and Organizational Psychology* 32, no. 2 (2023): 216–33; Joachim Waterschoot, Jolene van der Kaap-Deeder, and Maarten Vansteenkiste, "The Role of Confidence-Related Attentional Bias and Resilience in Restoring Thwarted Feelings of Confidence," *Motivation and Emotion* 44 (2020): 82–98.

139 **expectations to perform**: Nicolas Gillet et al., "The Effects of Job Demands and Organizational Resources through Psychological Need Satisfaction and Thwarting," *Spanish Journal of Psychology* 18 (2015): E28.

139 **monotony can also**: For more, see Kimberley Jane Bartholomew et al., "Job Pressure and Ill-Health in Physical Education Teachers: The Mediating Role of Psychological Need Thwarting," *Teaching and Teacher Education* 37 (2014): 101–7; Chris Giebe and Thomas Rigotti, "Tenets of Self-Determination Theory as a Mechanism Behind Challenge Demands: A Within-Person Study," *Journal of Managerial Psychology* 37, no. 5 (2022): 480–97.

139 **which can lead to worries**: Kimberley J. Bartholomew et al., "Psychological Need Thwarting in the Sport Context: Assessing the Darker Side of Athletic Experience," *Journal of Sport and Exercise Psychology* 33, no. 1 (2011): 75–102.

139 **unclear expectations and unexpected shifts**: Gillet et al., "Effects of Job Demands and Organizational Resources through Psychological Need Satisfaction and Thwarting," E28.

139 **setbacks**: Waterschoot, van der Kaap-Deeder, and Vansteenkiste, "Role of Confidence-Related Attentional Bias and Resilience in Restoring Thwarted Feelings of Confidence," 82–98.

139 **criticism**: Huyghebaert-Zouaghi et al., "Managerial Predictors and Motivational Outcomes of Workers' Psychological Need States Profiles," 216–33; Waterschoot, van der Kaap-Deeder, and Vansteenkiste, "Role of Confidence-Related Attentional Bias and Resilience in Restoring Thwarted Feelings of Confidence," 82–98.

139 **inferiority**: Bartholomew et al., "Psychological Need Thwarting in the Sport Context," 75–102.

139 **inadequate**: Bartholomew et al., "Psychological Need Thwarting in the Sport Context," 75–102.

139 **inferior, demotivated**: Jungwoo Ha, "Not Being Able to Verify One's Confidence: Negative Consequences of Thwarted Self-Promotion," *Academy of Management Proceedings* 2017, no. 1 (2017).

139 **we're a failure**: Marie-Christine Opdenakker, "Need-Supportive and Need-Thwarting Teacher Behavior: Their Importance to Boys' and Girls' Academic Engagement and Procrastination Behavior," *Frontiers in Psychology* 12 (2021): 628064.

Notes

139 **excessive self-focus**: Ha, "Not Being Able to Verify One's Confidence."
139 **unnecessary competitiveness**: Hui Fang et al., "Being Eager to Prove Oneself: U-Shaped Relationship between Confidence Frustration and Intrinsic Motivation in Another Activity," *Frontiers in Psychology* 8 (2017): 2123.
139 **paranoia**: Ha, "Not Being Able to Verify One's Confidence."
139 **negative self-talk**: Opdenakker, "Need-Supportive and Need-Thwarting Teacher Behavior," 628064.
139 **exhaustion**: Ricardo Cuevas-Campos et al., "Need Satisfaction and Need Thwarting in Physical Education and Intention to Be Physically Active," *Sustainability* 12, no. 18 (2020): 7312.
139 **burnout**: Bartholomew et al., "Job Pressure and Ill-Health in Physical Education Teachers," 101–7.
139 **in relationships, they lead to**: Meredith Rocchi et al., "Assessing Need-Supportive and Need-Thwarting Interpersonal Behaviours: The Interpersonal Behaviours Questionnaire (IBQ)," *Personality and Individual Differences* 104 (2017): 423–33.
140 **invigorating**: Naoki Miura et al., "Neural Evidence for the Intrinsic Value of Action as Motivation for Behavior," *Neuroscience* 352 (2017): 190–203.
140 **and important**: Dean A. Shepherd and Melissa S. Cardon, "Negative Emotional Reactions to Project Failure and the Self-Compassion to Learn from the Experience," *Journal of Management Studies* 46, no. 6 (2009): 923–49.
140 **a state of "flow"**: Judith M. Harackiewicz, Steven Abrahams, and Ruth Wageman, "Performance Evaluation and Intrinsic Motivation: The Effects of Evaluative Focus, Rewards, and Achievement Orientation," *Journal of Personality and Social Psychology* 53, no. 6 (1987): 1015.
140 **focused pursuit of your objective**: Maarten Vansteenkiste et al., "Autonomy and Relatedness among Chinese Sojourners and Applicants: Conflictual or Independent Predictors of Well-Being and Adjustment?," *Motivation and Emotion* 30 (2006): 273–82.
140 **motivated**: Judith M. Harackiewicz, Steven Abrahams, and Ruth Wageman, "Performance Evaluation and Intrinsic Motivation: The Effects of Evaluative Focus, Rewards, and Achievement Orientation," *Journal of Personality and Social Psychology* 53, no. 6 (1987): 1015.
140 **interested**: Ann K. Boggiano, Deborah S. Main, and Phyllis A. Katz, "Children's Preference for Challenge: The Role of Perceived Confidence And Control," *Journal of Personality and Social Psychology* 54, no. 1 (1988): 134.
140 **and energized**: Diego Vasconcellos et al., "Self-Determination Theory Applied to Physical Education: A Systematic Review and Meta-Analysis," *Journal of Educational Psychology* 112, no. 7 (2020): 1444.
140 **humans are wired**: Yu et al., "Self-Determined Motivation to Choose College Majors, Its Antecedents, and Outcomes," 132–50.
140 **Complex mental processes**: Konrad Lorenz, *The Waning of Humaneness* (London: Unwin Hyman, 1988); see also Ann Wilcock, "A Theory of the Human Need for Occupation," *Journal of Occupational Science* 1, no. 1 (1993): 17–24.
140 **our ancestors' survival**: Sven Schrader et al., "Cortext: A Columnar Model of Bottom-Up and Top-Down Processing in the Neocortex," *Neural Networks* 22, no. 8 (2009): 1055–70.
140 **"occupational beings"**: Wilcock, "Theory of the Human Need for Occupation," 17; see also Edward L. Deci and Richard M. Ryan, "The 'What' and 'Why' of Goal Pursuits: Human Needs and the Self-Determination of Behavior," *Psychological Inquiry* 11, no. 4 (2000): 227–68.

143 **projecting success externally**: Gordon L. Flett and Paul L. Hewitt, "Managing Perfectionism and the Excessive Striving That Undermines Flourishing: Implications for Leading the Perfect Life," in *Flourishing in Life, Work and Careers: Individual Wellbeing and Career Experiences*, ed. Ronald J. Burke, Kathryn M. Page, and Cary L. Cooper (Cheltenham, UK: Edward Elgar Publishing, 2015), 45–66.

143 **tend to fuse their self-worth**: Isabel Marten DiBartoloIsabel et al., "Shedding Light on the Relationship between Personal Standards and Psychopathology: The Case for Contingent Self-Worth," *Journal of Rational-Emotive and Cognitive-Behavior Therapy* 22 (2004): 237–50.

143 **More reactive**: Flett and Hewitt, "Managing Perfectionism and the Excessive Striving That Undermines Flourishing," 45–66.

143 **shortcomings as proof**: Jofel D. Umandap and Lota A. Teh, "Self-Compassion as a Mediator between Perfectionism and Personal Growth Initiative," *Psychological Studies* 65 (2020): 227–38.

143 **asking for help**: Flett and Hewitt, "Managing Perfectionism and the Excessive Striving That Undermines Flourishing," 45–66.

143 **recipe for self-criticism, guilt**: Susan M. Bögels and Warren Mansell, "Attention Processes in the Maintenance and Treatment of Social Phobia: Hypervigilance, Avoidance and Self-Focused Attention," *Clinical Psychology Review* 24, no. 7 (2004): 827–56.

143 **and shame**: Umandap and Teh, "Self-Compassion as a Mediator between Perfectionism and Personal Growth Initiative," 227–38.

143 **disrupted sleep**: Allison G. Harvey and Emmeline Greenall, "Catastrophic Worry in Primary Insomnia," *Journal of Behavior Therapy and Experimental Psychiatry* 34, no. 1 (2003): 11–23.

143 **thought to mostly affect high achievers**: Umandap and Teh, "Self-Compassion as a Mediator between Perfectionism and Personal Growth Initiative," 227–38.

143 **steady rise in perfectionism**: Thomas Curran and Andrew P. Hill, "Perfectionism Is Increasing over Time: A Meta-Analysis of Birth Cohort Differences from 1989 to 2016," *Psychological Bulletin* 145, no. 4 (2019): 410.

144 **imposter syndrome**: Kenneth T. Wang, Marina S. Sheveleva, and Tatiana M. Permyakova, "Imposter Syndrome among Russian Students: The Link between Perfectionism and Psychological Distress," *Personality and Individual Differences* 143 (2019): 1–6.

144 **avoid feeling inferior**: Adam Neufeld et al., "Why Do We Feel Like Intellectual Frauds? A Self-Determination Theory Perspective on the Impostor Phenomenon in Medical Students," *Teaching and Learning in Medicine* 35, no. 2 (2023): 180–92.

144 **most undermines our well-being**: Wang, Sheveleva, and Permyakova, "Imposter Syndrome among Russian Students," 1–6.

145 **we think *others* see us**: Mark R. Leary et al., "Self-Esteem as an Interpersonal Monitor: The Sociometer Hypothesis," *Journal of Personality and Social Psychology* 68, no. 3 (1995): 518; Mark R. Leary and Roy F. Baumeister, "The Nature and Function of Self-Esteem: Sociometer Theory," in *Advances in Experimental Social Psychology*, vol. 32, ed. Mark P. Zanna (San Francisco: Academic Press, 2000), 1–62.

145 **metaperception**: David A. Kenny, *Interpersonal Perception: A Social Relations Analysis* (New York: Guilford Press, 1994).

145 **feedback-rich environments**: Adam M. Mastroianni et al., "The Liking Gap in Groups and Teams," *Organizational Behavior and Human Decision Processes* 162 (2021): 109–22.

145 **than they actually do**: Jungwoo Ha, "The Impact of Thwarted Competence-Presentation on Turnover Intentions," *Academy of Management Proceedings* 2015, no. 1 (2015): https://doi.org/10.5465/ambpp.2015.14824abstract.

145 **habit of self-criticism**: Flett and Hewitt, "Managing Perfectionism and the Excessive Striving That Undermines Flourishing," 45–66.

146 **through the eyes of others**: Shmuel Ellis and Inbar Davidi, "After-Event Reviews: Drawing Lessons from Successful and Failed Experience," *Journal of Applied Psychology* 90, no. 5 (2005): 857.

146 **reflected best self exercise**: Laura Morgan Roberts et al., "Composing the Reflected Best-Self Portrait: Building Pathways for Becoming Extraordinary in Work Organizations," *Academy of Management Review* 30, no. 4 (2005): 712–36.

146 **boosts subjective confidence**: Noelle Baird, Jennifer L. Robertson, and Matthew J. W. McLarnon, "Looking in the Mirror: Including the Reflected Best Self Exercise in Management Curricula to Increase Students' Interview Self-Efficacy," *Academy of Management Learning & Education* 22, no. 4 (2023): 662–80.

146 **renews psychological resources and increases loving kindness**: Gretchen Spreitzer, John Paul Stephens, and David Sweetman, "The Reflected Best Self Field Experiment with Adolescent Leaders: Exploring the Psychological Resources Associated with Feedback Source and Valence," *Journal of Positive Psychology* 4, no. 5 (2009): 331–48.

147 **a diverse mix**: Spreitzer, Stephens, and Sweetman, "Reflected Best Self Field Experiment with Adolescent Leaders," 331–48.

147 Spreitzer, Stephens, and Sweetman, "Reflected Best Self Field Experiment with Adolescent Leaders," 331–48.

147 ***more* positive emotions**: Spreitzer, Stephens, and Sweetman, "Reflected Best Self Field Experiment with Adolescent Leaders," 331–48.

147 **Gathering a broad portrait**: Neufeld et al., "Why Do We Feel Like Intellectual Frauds?," 180–92.

150 **starve her doubt dragon**: Marilyn V. Whitman and Kristen K. Shanine, "Revisiting the Impostor Phenomenon: How Individuals Cope with Feelings of Being in over Their Heads," in *The Role of the Economic Crisis on Occupational Stress and Well Being*, ed. Pamela L. Perrewé, Jonathan R. B. Halbesleben, and Christopher C. Rosen (Bingley, UK: Emerald Group, 2012), 177–212.

150 **regular confidence crafting**: Wang, Sheveleva, and Permyakova, "Imposter Syndrome among Russian Students," 1–6; Neufeld et al., "Why Do We Feel Like Intellectual Frauds?," 180–92.

151 **conditional acceptance**: A. Macedo et al., "Conditional Acceptance as a Distinct Feature of Socially-Prescribed Perfectionism: A Study in Portuguese Pregnant Women," *European Psychiatry* 24, no. S1 (2009): 1; Flett and Hewitt, "Managing Perfectionism and the Excessive Striving That Undermines Flourishing," 45–66.

151 **A related thinking style**: Roz Shafran, Anna Coughtrey, and Radha Kothari, "New Frontiers in the Treatment of Perfectionism," *International Journal of Cognitive Therapy* 9, no. 2 (2016): 156–70.

151 **I'm a total failure**: Sarah J. Egan et al., "The Role of Dichotomous Thinking and Rigidity in Perfectionism," *Behaviour Research and Therapy* 45, no. 8 (2007): 1813–22.

Notes

155 **the more success we've had**: Riccardo Leoncini, "How to Learn from Failure. Organizational Creativity, Learning, Innovation and the Benefit of Failure," *Rutgers Business Review* 2, no. 1 (2017): 98–104.

157 **A drive to protect ourselves**: Henry A. Murray, *Explorations in Personality* (New York: Oxford University Press, 1938).

157 **overreacting to**: Murray, *Explorations in Personality*.

157 **downplaying mistakes**: Murray, *Explorations in Personality*.

157 **seeking superiority**: Kristin D. Neff, Ya-Ping Hsieh, and Kullaya Dejitterat, "Self-Compassion, Achievement Goals, and Coping with Academic Failure," *Self and Identity* 4, no. 3 (2005): 263.

157 **focus on ourselves**: Avi Kaplan and Martin L. Maehr, "Achievement Goals and Student Well-Being," *Contemporary Educational Psychology* 24, no. 4 (1999): 330–58.

157 **or our performance**: Bögels and Mansell, "Attention Processes in the Maintenance and Treatment of Social Phobia," 827–56.

157 **Compulsive focus**: Maarten Vansteenkiste and Richard M. Ryan, "On Psychological Growth and Vulnerability: Basic Psychological Need Satisfaction and Need Frustration as a Unifying Principle," *Journal of Psychotherapy Integration* 23, no. 3 (2013): 263; Kim Tolentino, Tucker Readdy, and Johannes Raabe, "'No Days Off': Using Self-Determination Theory to Better Understand Workaholism in National Collegiate Athletic Association Division I Coaches," *Journal of Clinical Sport Psychology* 18, no. 2 (2022): 251–69.

157 **Excessive competition**: Fang et al., "Being Eager to Prove Oneself," 2123.

157 **Obsessive passion**: Daniel Lalande et al., "Obsessive Passion: A Compensatory Response to Unsatisfied Needs," *Journal of Personality* 85, no. 2 (2017): 163–78.

157 **workaholism**: Tolentino, Readdy, and Raabe, "No Days Off," 251–69.

157 **Frequent self-criticism**: Opdenakker, "Need-Supportive and Need-Thwarting Teacher Behavior," 628064.

157 **mean we're a failure**: Vansteenkiste and Ryan, "On Psychological Growth and Vulnerability," 263.

157 **Paranoia**: Ha, "Not Being Able to Verify One's Confidence."

157 **Seeking constant praise**: Mireia Las Heras and D. T. Hall, "Integration of Career and Life," *Handbook on Women in Business and Management*, ed. Diana Bilimoria and Sandy Kristin Piderit (Cheltenham, UK: Edward Elgar Publishing, 2007), 178–205; Valerie Good et al., "A Self-Determination Theory-Based Meta-Analysis on the Differential Effects of Intrinsic and Extrinsic Motivation on Salesperson Performance," *Journal of the Academy of Marketing Science* 50, no. 3 (2022): 586–614.

157 **obsession with impressing others; Believing our worth**: Joke Verstuyf et al., "Motivational Dynamics of Eating Regulation: A Self-Determination Theory Perspective," *International Journal of Behavioral Nutrition and Physical Activity* 9, no. 1 (2012): 1–16.

158 **Self-handicapping behaviors**: Verstuyf et al., "Motivational Dynamics of Eating Regulation," 1–16.

Chapter 10: Crafting Choice

159 **to whom he's been compared**: Paul Gray, "The Magic of Harry Potter," TIME, December 25, 2000, https://content.time.com/time/subscriber/article/0,33009,998844-2,00.html.

159 **After studying law**: Ian Traynor, "Slobodan Milosevic," Guardian, March 12, 2006, https://www.theguardian.com/news/2006/mar/13/guardianobituaries.warcrimes.

Notes

159 **dough-faced**: Laura Secor, "Empty Vessel," Nation, April 19, 2004, https://www.thenation.com/article/archive/empty-vessel.

159 **big-eared**: Blaine Harden, "The Milosevic Generation," *New York Times Magazine*, August 29, 1999, https://www.nytimes.com/1999/08/29/magazine/the-milosevic-generation.html.

159 **wooden**: Harden, "Milosevic Generation."

159 **unremarkable**: Christopher Hitchens, "No Sympathy for Slobo," Slate, March 13, 2006, https://slate.com/news-and-politics/2006/03/slobodan-milosevic-resentful-nonentity-blood thirsty-dictator.html.

159 **rode a wave of xenophobic nationalism**: John B. Allcock, "Slobodan Milošević," *Encyclopedia Britannica*, last updated August 25, 2024, https://www.britannica.com/biography/Slobodan-Milosevic.

159 **"Butcher of the Balkans"**: Steve York, dir., *Bringing Down a Dictator* (Washington, DC: International Center on Nonviolent Conflict, 2002), https://www.nonviolent-conflict.org/bringing-dictator-english.

159 **entered into bloody conflict**: Roger Cohen, "THE WORLD: PAST REASON; Yes, Blood Stains the Balkans. No, It's Not Just Fate," *New York Times*, October 4, 1998, https://www.nytimes.com/1998/10/04/weekinreview/the-world-past-reason-yes-blood-stains-the-balkans-no-it-s-not-just-fate.html.

160 **torching villages**: US Department of State, "Ethnic Cleansing in Kosovo: An Accounting," State Department Report, December 1999, US Department of State Archive, accessed September 11, 20224, https://1997-2001.state.gov/global/human_rights/kosovoii/homepage.html.

160 **building concentration camps**: Remembering Srebrenica, "Concentration Camps," accessed September 11, 2024, https://srebrenica.org.uk/what-happened/history/concentration-camps.

160 **raping tens of thousands**: Maria B. Olujic, "Coercion and Torture in Former Yugoslavia," Cultural Survival, March 19, 2010, https://www.culturalsurvival.org/publications/cultural-survival-quarterly/coercion-and-torture-former-yugoslavia.

160 **125,000 deaths**: Before the ceasefire, ten thousand people had been killed in Croatia, a number that would have certainly have been higher without UN peacekeeping troops (History.com Editors, "Former Yugoslav President Slobodan Milosevic Goes on Trial for War Crimes," History, A&E Television Netowrks, last updated February 9, 2024, https://www.history.com/this-day-in-history/milosevic-goes-on-trial-for-war-crimes); one hundred thousand in Bosnia (United Nations International Criminal Tribunal for the Former Yugoslavia, "The Conflicts," United Nations International Residual Mechanism for Criminal Tribunals, accessed September 11, 2024, https://www.icty.org/en/about/what-former-yugoslavia/conflicts), including eight thousand Muslim men and boys in a Bosnian UN safe zone ("Timeline: Serbia, 20 Years since Milosevic Came to Power," Reuters, September 20, 2010, http://www.reuters.com/article/idUSTRE68T12W); and at least thirteen thousand in Kosovo (Valerie Plesch, "'They Took Him from My Hands': Kosovo War Massacre Remembered," Al Jazeera, April 28, 2019, https://www.aljazeera.com/news/2019/4/28/they-took-him-from-my-hands-kosovo-war-massacre-remembered).

160 **Displacement of three million**: In Bosnia, two million people (half the population) became refugees (UN International Criminal Tribunal for the Former Yugoslavia, "Conflicts"); in Kosovo, one and a half million were expelled from their homes (80 percent of Kosovo's population [Human Rights Watch, "Under Orders: War Crimes in Kosovo," Executive Summary,

Notes

October 26, 2001, https://www.hrw.org/report/2001/10/26/under-orders/war-crimes-kosovo]) (US Department of State, "Ethnic Cleansing in Kosovo").

160 **life was bleak**: Srdja Popovic and Matthew Miller, *Blueprint for Revolution: How to Use Rice Pudding, Lego Men, and Other Nonviolent Techniques to Galvanize Communities, Overthrow Dictators, or Simply Change the World* (New York: Random House, 2015).

160 **seven million citizens**: https://www.macrotrends.net/global-metrics/countries/SRB/serbia/population#google_vignette.

160 **Milošević murdered political adversaries**: Stone Soup Leadership Institute, "A Blueprint for Climate Revolution: Srda Popovic," SustainabilityIsFun, accessed September 11, 2024, https://sustainabilityisfun.net/wp-content/uploads/2021/04/story-srdja-popovic.pdf.

160 **into unprecedented poverty**: Hitchens, "No Sympathy for Slobo."

160 **The second son of two TV reporters**: Jon Henley, "Meet Srdja Popovic, the Secret Architect of Global Revolution," Guardian, March 8, 2015, https://www.theguardian.com/world/2015/mar/08/srdja-popovic-revolution-serbian-activist-protest.

160 **typical college students**: Popovic and Miller, *Blueprint for Revolution*.

160 **everything he stood for**: Henley, "Meet Srdja Popovic."

161 **regime-appointed deans**: RefWorld: Global Law & Policy Database, "Deepening Authoritarianism in Serbia: The Purge of the Universities," Human Rights Watch, January 1, 1999, UNHCR, https://www.refworld.org/reference/countryrep/hrw/1999/en/96731.

161 **surrender or take a stand**: Roger Cohen, "Who Really Brought Down Milosevic," *New York Times Magazine*, November 26, 2000, https://www.nytimes.com/2000/11/26/magazine/who-really-brought-down-milosevic.html.

161 **Srdja's friend Duda**: Popovic and Miller, *Blueprint for Revolution*.

161 **flyers with pithy slogans**: York, *Bringing Down a Dictator*.

161 **"potential police informants"**: Popovic and Miller, *Blueprint for Revolution*.

162 **make our own choices**: Edward L. Deci and Richard M. Ryan, "The Importance of Autonomy for Development and Well-Being," in *Self-Regulation and Autonomy: Social and Developmental Dimensions of Human Conduct*, ed. Bryan W. Sokol, Frederick M. E. Grouzet, and Ulrich Müller (New York: Cambridge University Press, 2013), 19–46.

162 **express ourselves**: Nele Laporte et al., "Adolescents as Active Managers of Their Own Psychological Needs: The Role of Psychological Need Crafting in Adolescents' Mental Health," *Journal of Adolescence* 88 (2021): 67–83; Kennon M. Sheldon et al., "Trait Self and True Self: Cross-Role Variation in the Big-Five Personality Traits and Its Relations with Psychological Authenticity and Subjective Well-Being," *Journal of Personality and Social Psychology* 73, no. 6 (1997): 1380.

162 **values, interests**: Marie-Christine Opdenakker, "Need-Supportive and Need-Thwarting Teacher Behavior: Their Importance to Boys' and Girls' Academic Engagement and Procrastination Behavior," *Frontiers in Psychology* 12 (2021): 628064.

162 **and needs**: Guy Roth, "Antecedents and Outcomes of Teachers' Autonomous Motivation: A Self-Determination Theory Analysis," in *Teacher Motivation: Theory and Practice*, ed. Paul W. Richardson, Stuart A. Karabenick, and Helen M. G. Watt, (New York: Routledge, 2014), 36–51.

162 **every revolution and quest for freedom**: Edward L. Deci and Richard M. Ryan, "Motivation, Personality, and Development within Embedded Social Contexts: An Overview of Self-Determination Theory," in *The Oxford Handbook of Human Motivation*, ed. Richar M. Ryan (Oxford: Oxford University Press, 2012), 85–107.

Notes

163 **How do I *really* feel**: John P. Meyer et al., "Motivational Mindsets versus Reasons for Action: Implications for the Dimensionality Debate in Self-Determination Theory," *Motivation and Emotion* 46, no. 4 (2022): 486–507.

163 **deep sense of self**: Sheldon et al., "Trait Self and True Self," 1380.

163 **because we *want* to**: Anne C. Holding et al., "Sacrifice—But at What Price? A Longitudinal Study of Young Adults' Sacrifice of Basic Psychological Needs in Pursuit of Career Goals," *Motivation and Emotion* 44 (2020): 99–115.

163 **architect of your life**: John Wild, "Authentic Existence," *Ethics* 75, no. 4 (1965): 227–39; see also Kennon M. Sheldon et al., "What Is Satisfying about Satisfying Events? Testing 10 Candidate Psychological Needs," *Journal of Personality and Social Psychology* 80, no. 2 (2001): 325.

163 **inner awareness**: Maarten Vansteenkiste, Christopher P. Niemiec, and Bart Soenens, "The Development of the Five Mini-Theories of Self-Determination Theory: An Historical Overview, Emerging Trends, and Future Directions," in *The Decade Ahead: Theoretical Perspectives on Motivation and Achievement*, ed. Timothy C. Urdan and Stuart A. Karabenick (Bingley, UK: Emerald, 2010), 105–65.

163 **alignment**: Deci and Ryan, "Importance of Autonomy for Development and Well-Being," 19–46.

163 **conserves mental energy**: Maarten Vansteenkiste and Richard M. Ryan, "On Psychological Growth and Vulnerability: Basic Psychological Need Satisfaction and Need Frustration as a Unifying Principle," *Journal of Psychotherapy Integration* 23, no. 3 (2013): 263.

163 **powers well-being**: Specifically more positive emotions and less reactivity: Richard M. Ryan et al., "We Know This Much Is (Meta-Analytically) True: A Meta-Review of Meta-Analytic Findings Evaluating Self-Determination Theory," *Psychological Bulletin* 148, no. 11–12 (2022): 813.

163 **more engaged**: Avi Assor, Haya Kaplan, and Guy Roth, "Choice Is Good, but Relevance Is Excellent: Autonomy-Enhancing and Suppressing Teacher Behaviours Predicting Students' Engagement in Schoolwork," *British Journal of Educational Psychology* 72, no. 2 (2002): 261–78.

163 **creative**: Deci and Ryan, "Importance of Autonomy for Development and Well-Being," 19–46.

163 **fulfilled**: Erika A. Patall, Harris Cooper, and Jorgianne Civey Robinson, "The Effects of Choice on Intrinsic Motivation and Related Outcomes: A Meta-Analysis of Research Findings," *Psychological Bulletin* 134, no. 2 (2008): 270.

163 **service to others**: Deci and Ryan, "Importance of Autonomy for Development and Well-Being," 19–46.

163 **stress**: Pamela K. Smith and Wilhelm Hofmann, "Power in Everyday Life," *Proceedings of the National Academy of Sciences* 113, no. 36 (2016): 10043–48.

163 **unhappiness**: Luke Felton and Sophia Jowett, "On Understanding the Role of Need Thwarting in the Association between Athlete Attachment and Well/Ill-Being," *Scandinavian Journal of Medicine & Science in Sports* 25, no. 2 (2015): 289–98.

163 **anxiety and even depression**: See Ozge Tayfur, "The Antecedents and Consequences of Learned Helplessness in Work Life," *Information Management and Business Review* 4, no. 7 (2012): 417–27.

163 **adopt shadow goals**: For example, Rémi Radel et al., "Restoration Process of the Need for Autonomy: The Early Alarm Stage," *Journal of Personality and Social Psychology* 101, no. 5 (2011): 919.

164 **boredom**: Ricardo Cuevas-Campos et al., "Need Satisfaction and Need Thwarting in Physical Education and Intention to Be Physically Active," *Sustainability* 12, no. 18 (2020): 7312.
164 **apathy**: Hui Fang et al., "The Spillover Effect of Autonomy Frustration on Human Motivation and Its Electrophysiological Representation," *Frontiers in Human Neuroscience* 14 (2020): 134.
164 **exhaustion**: Cuevas-Campos et al., "Need Satisfaction and Need Thwarting in Physical Education and Intention to be Physically Active," 7312.
164 **stifle participation**: Blake E. Ashforth, "The Experience of Powerlessness in Organizations," *Organizational Behavior and Human Decision Processes* 43, no. 2 (1989): 207–42.
164 **silence voices**: Özge Tayfur, "Çalışma hayatında öğrenilmiş çaresizlik ve tükenmişliğin nedenleri ve sonuçları üzerine bir çalışma" (2011). Unpublished manuscript.
164 **stop growth**: Ashforth, "Experience of Powerlessness in Organizations," 207–42.
164 **declines in performance**: Tayfur, "Antecedents and Consequences of Learned Helplessness in Work Life," 417–27.
164 **mistakes**: Constance R. Campbell and Mark J. Martinko, "An Integrative Attributional Perspective of Empowerment and Learned Helplessness: A Multimethod Field Study," *Journal of Management* 24, no. 2 (1998): 173–200.
164 **absenteeism, and even unethical conduct**: Tayfur, "Çalışma hayatında öğrenilmiş çaresizlik ve tükenmişliğin nedenleri ve sonuçları üzerine bir çalışma."
164 **Tens of thousands**: Mark Schapiro, "Serbia's Lost Generation," *Mother Jones*, September/October 1999, https://www.motherjones.com/politics/1999/09/serbias-lost-generation.
164 **only escape was mental**: York, *Bringing Down a Dictator*.
164 **denial, acceptance, or apathy**: Radel et al., "Restoration Process of the Need for Autonomy," 919.
164 **belting out such lyrics as**: Henley, "Meet Srdja Popovic."
165 **the smiling barrel**: Popovic and Miller, *Blueprint for Revolution*.
166 **"ideological commissar"**: Cohen, "Who Really Brought Down Milosevic."
166 **choice support**: Wendy S. Grolnick, Richard M. Ryan, and Edward L. Deci, "Inner Resources for School Achievement: Motivational Mediators of Children's Perceptions of Their Parents," *Journal of Educational Psychology* 83, no. 4 (1991): 508; Catherine F. Ratelle, Karine Simard, and Frédéric Guay, "University Students' Subjective Well-Being: The Role of Autonomy Support from Parents, Friends, and the Romantic Partner," *Journal of Happiness Studies* 14 (2013): 893–910.
167 **bosses**: Ratelle, Simard, and Guay, "University Students' Subjective Well-Being," 893–910.
167 **educators**: Johnmarshall Reeve, Hye-Ryen Jang, and Hyungshim Jang, "Personality-Based Antecedents of Teachers' Autonomy-Supportive and Controlling Motivating Styles," *Learning and Individual Differences* 62 (2018): 12–22.
167 **coaches**: María Sol Álvarez et al., "Coach Autonomy Support and Quality of Sport Engagement in Young Soccer Players," *Spanish Journal of Psychology* 12, no. 1 (2009): 138–48.
167 **therapists**: David C. Zuroff et al., "Therapist's Autonomy Support and Patient's Self-Criticism Predict Motivation during Brief Treatments for Depression," *Journal of Social and Clinical Psychology* 31, no. 9 (2012): 903–32.
167 **parents**: Maarten Vansteenkiste et al., "Longitudinal Associations between Adolescent Perceived Degree and Style of Parental Prohibition and Internalization and Defiance," *Developmental Psychology* 50, no. 1 (2014): 229.

Notes

167 **siblings**: Élodie C. Audet et al., "A Remarkable Alliance: Sibling Autonomy Support and Goal Progress in Emerging Adulthood," *Family Relations* 70, no. 5 (2021): 1571–82.

167 **friends**: Ratelle, Simard, and Guay, "University Students' Subjective Well-Being," 893–910.

167 **partners**: Netta Weinstein et al., "Autonomy Support and Diastolic Blood Pressure: Long Term Effects and Conflict Navigation in Romantic Relationships," *Motivation and Emotion* 40 (2016): 212–25.

167 **find new agency**: Yu-Lan Su and Johnmarshall Reeve, "A Meta-Analysis of the Effectiveness of Intervention Programs Designed to Support Autonomy," *Educational Psychology Review* 23 (2011): 159–88.

167 **greater need satisfaction**: Sara Kindt et al., "Helping Motivation and Well-Being of Chronic Pain Couples: A Daily Diary Study," *Pain* 157, no. 7 (2016): 1551–62; Kimberley J. Bartholomew et al., "Self-Determination Theory and Diminished Functioning: The Role of Interpersonal Control and Psychological Need Thwarting," *Personality and Social Psychology Bulletin* 37, no. 11 (2011): 1459–73.

167 **motivation**: Liang Meng and Qingguo Ma, "Live as We Choose: The Role of Autonomy Support in Facilitating Intrinsic Motivation," *International Journal of Psychophysiology* 98, no. 3 (2015): 441–47.

167 **growth**: Also known as agentic engagement: Johnmarshall Reeve, "Giving and Summoning Autonomy Support in Hierarchical Relationships," *Social and Personality Psychology Compass* 9, no. 8 (2015): 406–18.

167 **performance**: Naniki Mokgata, Leoni van der Vaart, and Leon T. de Beer, "Autonomy-Supportive Agents: Whose Support Matters Most, and How Does It Unfold in the Workplace?," *Current Psychology* 42, no. 27 (2023): 23931–46.

167 **choice-supportive behaviors**: Mark Muraven, Marylene Gagné, and Heather Rosman, "Helpful Self-Control: Autonomy Support, Vitality, and Depletion," *Journal of Experimental Social Psychology* 44, no. 3 (2008): 573–85.

167 **running a haunted house**: Popovic and Miller, *Blueprint for Revolution*.

168 **"This is THE year..."**: Popovic and Miller, *Blueprint for Revolution*.

168 **moved the elections up**: RadioFreeEurope/RadioLiberty, "Timeline: The Political Career of Slobodan Milosevic," Regions: Montenegro, March 13, 2006, https://www.rferl.org/a/1066641.html.

168 **untainted by problematic political ties**: Audrey Helfant Budding, "'The Man Who Overthrew Milosevic': Vojislav Kostunica, One Year Later," *Fletcher Forum of World Affairs* 26, no. 1 (2002): 159–65.

168 **Thirty thousand volunteers**: Yhork, *Bringing Down a Dictator*.

168 **Nearly 90 percent**: Stone Soup Leadership Institute, "Blueprint for Climate Revolution."

168 **series of civil actions**: RadioFreeEurope/RadioLiberty, "Timeline."

169 **Their own kids**: York, *Bringing Down a Dictator*.

169 **International Criminal Court for war crimes**: RadioFreeEurope/RadioLiberty, "Timeline."

169 **found dead in his cell**: "Timeline: Serbia, 20 Years since Milosevic Came to Power."

171 **with the objective reality**: Smith and Hofmann, "Power in Everyday Life," 10043–48.

171 **motivated**: Donald C. Pelz and Frank M. Andrews, *Scientists in Organizations: Productive Climates for Research and Development* (New York: Wiley, 1966).

Notes

171 **ambitious, high achievers**: Ashforth, "Experience of Powerlessness in Organizations," 207–42; Tayfur, "Antecedents and Consequences of Learned Helplessness in Work Life," 417–27; Martinko and Gardner (1982).

171 **new, creative ways**: Lisa Legault et al., "Assisted versus Asserted Autonomy Satisfaction: Their Unique Associations with Wellbeing, Integration of Experience, and Conflict Negotiation," *Motivation and Emotion* 41 (2017): 1–21.

173 **improves need satisfaction in others**: Jessica De Bloom et al., "An Identity-Based Integrative Needs Model of Crafting: Crafting within and across Life Domains," *Journal of Applied Psychology* 105, no. 12 (2020): 1423.

174 **what "good" professionals did**: Richard M. Ryan et al., "Building a Science of Motivated Persons: Self-Determination Theory's Empirical Approach to Human Experience and the Regulation of Behavior," *Motivation Science* 7, no. 2 (2021): 97.

174 **others' expectations over his own needs**: This study uses the term *conformity* (Abhijeet K. Vadera and Michael G. Pratt, "Love, Hate, Ambivalence, or Indifference? A Conceptual Examination of Workplace Crimes and Organizational Identification," *Organization Science* 24, no. 1 (2013): 172–88.

174 **our identity**: George Smith, "Self, Self-Concept, and Identity," in *Handbook of Self and Identity*, ed. Mark R. Leary and June Price Tangney, 2nd ed. (New York: Guilford Press, 2012), 69–104.

174 **social support and shared purpose**: Lakshmi Ramarajan, "Past, Present and Future Research on Multiple Identities: Toward an Intrapersonal Network Approach," *Academy of Management Annals* 8, no. 1 (2014): 589–659.

174 **overidentification**: Lorenzo Avanzi et al., "The Downside of Organizational Identification: Relations between Identification, Workaholism and Well-Being," *Work & Stress* 26, no. 3 (2012): 289–307.

174 **is detrimental**: Michael Ahearne et al., "It's a Matter of Congruence: How Interpersonal Identification between Sales Managers and Salespersons Shapes Sales Success," *Journal of the Academy of Marketing Science* 41 (2013): 625–48.

174 **all too common**: Glen E. Kreiner, Elaine C. Hollensbe, and Mathew L. Sheep, "Where Is the "Me" among the "We"? Identity Work and the Search for Optimal Balance," *Academy of Management Journal* 49, no. 5 (2006): 1031–57.

174 **even encourage it**: Mats Alvesson and Hugh Willmott, "Identity Regulation as Organizational Control: Producing the Appropriate Individual," *Journal of Management Studies* 39, no. 5 (2002): 619–44.

174 **"a pathology"**: Vadera and Pratt, "Love, Hate, Ambivalence, or Indifference?," 172–88.

174 **workaholism, poor well-being**: Avanzi et al., "Downside of Organizational Identification," 289–307.

174 **loneliness**: Peleg Dor-Haim and Izhar Oplatka, "School Principal's Perception of Loneliness: A Career Stage Perspective," *Journal of Educational Administration and History* 52, no. 2 (2020): 211–27.

175 **when our job becomes our identity**: Lauren L. Mitchell, Isabel A. Frazier, and Nina A. Sayer, "Identity Disruption and Its Association with Mental Health among Veterans with Reintegration Difficulty," *Developmental Psychology* 56, no. 11 (2020): 2152.

175 **means hating yourself**: Janna Koretz, "What Happens When Your Career Becomes Your Whole Identity," *Harvard Business Review*, December 26, 2019, https://hbr.org/2019/12/what-happens-when-your-career-becomes-your-whole-identity.

Notes

175 **decreased creativity, smart risk-taking**: Vadera and Pratt, "Love, Hate, Ambivalence, or Indifference?," 172–88.

175 **performance**: Ahearne et al., "It's a Matter of Congruence," 625–48.

175 **conflict, resistance to change**: For a review, see Samantha Conroy et al., "Where There Is Light, There Is Dark: A Review of the Detrimental Outcomes of High Organizational Identification," *Journal of Organizational Behavior* 38, no. 2 (2017): 184–203.

175 **unethical behavior**: Vadera and Pratt, "Love, Hate, Ambivalence, or Indifference?," 172–88. For a review, see Conroy et al., "Where There Is Light, There Is Dark," 184–203.

175 **suppress his authentic desires**: Sheldon et al., "Trait Self and True Self," 1380.

175 **a more balanced identity**: Anne Carolyn Muscat, "Elite Athletes' Experiences of Identity Changes during a Career-Ending Injury: An Interpretive Description" (PhD diss., University of British Columbia, 2010).

175 **turning down requests**: Jingyi Lu, Qingwen Fang, and Tian Qiu, "Rejecters Overestimate the Negative Consequences They Will Face from Refusal," *Journal of Experimental Psychology: Applied* 29, no. 2 (2022).

176 **"instead of yes"**: Arthur C. Brooks, "Overwhelmed? Just Say 'No,'" *Atlantic*, February 29, 2024, https://www.theatlantic.com/ideas/archive/2024/02/saying-no-science-happiness/677579.

176 **cared about outside his job**: This is a construct inspired by research on leisure crafting and off-job crafting (Justin M. Berg, Adam M. Grant, and Victoria Johnson, "When Callings Are Calling: Crafting Work and Leisure in Pursuit of Unanswered Occupational Callings," *Organization Science* 21, no. 5 (2010): 973–94; MiikaKujanpää et al., "Needs-Based Off-Job Crafting across Different Life Domains and Contexts: Testing a Novel Conceptual and Measurement Approach," *Frontiers in Psychology* 13 (2022): 959296).

176 **much fuller and richer**: Miika Kujanpää et al., "The Forgotten Ones: Crafting for Meaning and for Affiliation in the Context of Finnish and Japanese Employees' Off-Job Lives," *Frontiers in Psychology* 12 (2021): 682479.

176 **of a determined soul**: Ella Wheeler Wilcox, *Custer and Other Poems* (Chicago: W. B. Conkey, 1896).

179 **Build a balanced identity**: Glen E. Kreiner, Elaine C. Hollensbe, and Mathew L. Sheep, "Where Is the "Me" among the "We"? Identity Work and the Search for Optimal Balance," *Academy of Management Journal* 49, no. 5 (2006): 1031–57.

179 **need-crafting leisure activities**: Howard E. A. Tinsley and Barbara D. Eldredge, "Psychological Benefits of Leisure Participation: A Taxonomy of Leisure Activities Based on Their Need-Gratifying Properties," *Journal of Counseling Psychology* 42, no. 2 (1995): 123. Note that the article calls these needs self-expression, challenge, and affiliation, respectively, and measures them slightly differently. But conceptually, they are very similar.

179 **blindly defy expectations or rules**: Stijn Van Petegem et al., "Rebels with a Cause? Adolescent Defiance from the Perspective of Reactance Theory and Self-Determination Theory," *Child Development* 86, no. 3 (2015): 903–18; Vansteenkiste and Ryan, "On Psychological Growth and Vulnerability," 263; Richard Koestner, "Reaching One's Personal Goals: A Motivational Perspective Focused on Autonomy," *Canadian Psychology/Psychologie Canadienne* 49, no. 1 (2008): 60.

179 **refusing to follow social norms**: Stijn Van Petegem et al., "On the Association between Adolescent Autonomy and Psychosocial Functioning: Examining Decisional Independence from a Self-Determination Theory Perspective," *Developmental Psychology* 48, no. 1 (2012): 76.

Notes

179 **Rejecting authority**: Vansteenkiste and Ryan, "On Psychological Growth and Vulnerability," 263.

179 **the opposite of what's expected**: Vansteenkiste and Ryan, "On Psychological Growth and Vulnerability," 263; Maarten Vansteenkiste et al., "Examining Correlates of Game-to-Game Variation in Volleyball Players' Achievement Goal Pursuit and Underlying Autonomous and Controlling Reasons," *Journal of Sport and Exercise Psychology* 36, no. 2 (2014): 131–45.

179 **Passive aggressiveness**: Julien Bureau et al., "Promoting Autonomy to Reduce Employee Deviance: The Mediating Role of Identified Motivation," *International Journal of Business and Management* 13, no. 5 (2018): 61–71.

179 **Lying or cheating**: Yaniv Kanat-Maymon et al., "The Role of Basic Need Fulfillment in Academic Dishonesty: A Self-Determination Theory Perspective," *Contemporary Educational Psychology* 43 (2015): 1–9; Julien S. Bureau et al., "Investigating How Autonomy-Supportive Teaching Moderates the Relation between Student Honesty and Premeditated Cheating," *British Journal of Educational Psychology* 92, no. 1 (2022): 175–93.

180 **A drive to exercise control**: Richard M. Ryan, Veronika Huta, and Edward L. Deci, "Living Well: A Self-Determination Theory Perspective on Eudaimonia," *Journal of Happiness Studies* 9 (2008): 139–70. Added to aspirations index by Richard M. Ryan et al., "The American Dream in Russia: Extrinsic Aspirations and Well-Being in Two Cultures," *Personality and Social Psychology Bulletin* 25, no. 12 (1999): 1509–24.

180 **Using coercion**: Henry A. Murray, *Explorations in Personality* (New York: Oxford University Press, 1938).

180 **Hostility**: See Tayfur, "Antecedents and Consequences of Learned Helplessness in Work Life," 417–27.

180 **aggression**: Mireille Joussemet et al., "A Longitudinal Study of the Relationship of Maternal Autonomy Support to Children's Adjustment and Achievement in School," *Journal of Personality* 73, no. 5 (2005): 1215–36.

180 **bullying**: Kyriaki Fousiani et al., "Perceived Parenting and Adolescent Cyber-Bullying: Examining the Intervening Role of Autonomy and Relatedness Need Satisfaction, Empathic Concern and Recognition of Humanness," *Journal of Child and Family Studies* 25 (2016): 2120–29.

180 **order within ourselves or our surroundings**: Edward L. Deci and Richard M. Ryan, "The 'What' and 'Why' of Goal Pursuits: Human Needs and the Self-Determination of Behavior," *Psychological Inquiry* 11, no. 4 (2000): 227–68; Murray, *Explorations in Personality*.

180 **diet, exercise**: Deci and Ryan, "'What' and 'Why' of Goal Pursuits," 227–68.

180 **Apathy**: See Tayfur, "Antecedents and Consequences of Learned Helplessness in Work Life," 417–27.

180 **amotivation**: Deci and Ryan, "'What' and 'Why' of Goal Pursuits," 227–68; Shi Yu, Chantal Levesque-Bristol, and Yukiko Maeda, "General Need for Autonomy and Subjective Well-Being: A Meta-Analysis of Studies in the US and East Asia," *Journal of Happiness Studies* 19 (2018): 1863–82.

180 **Silence**: Campbell and Martinko, "Integrative Attributional Perspective of Empowerment and Learned Helplessness," 173–200.

180 **passivity**: Tayfur, "Antecedents and Consequences of Learned Helplessness in Work Life," 417–27.

Notes

180 **disengagement**: Ashforth, "Experience of Powerlessness in Organizations," 207–42.

180 **Ignoring or denying**: Radel et al., "Restoration Process of the Need for Autonomy," 919.

180 **Adopting beliefs, emotions, behaviors**: Tiphaine Huyghebaert et al., "Investigating the Longitudinal Effects of Surface Acting on Managers' Functioning through Psychological Needs," *Journal of Occupational Health Psychology* 23, no. 2 (2018): 207.

180 **Suppressing our real feelings**: Joke Verstuyf et al., "Motivational Dynamics of Eating Regulation: A Self-Determination Theory Perspective," *International Journal of Behavioral Nutrition and Physical Activity* 9, no. 1 (2012): 1–16.

180 **Self-medication**: Laporte et al., "Adolescents as Active Managers of Their Own Psychological Needs," 67–83; Verstuyf et al., "Motivational Dynamics of Eating Regulation," 1–16.

180 **Doing what others want**: Mohsin Bashir et al., "The Mediating Role of Psychological Need Thwarting in the Relationship between Compulsory Citizenship Behavior and Psychological Withdrawal," *Frontiers in Psychology* 10 (2019): 2595.

180 **Supporting controlling institutions**: Aaron C. Kay et al., "Compensatory Control: Achieving Order through the Mind, Our Institutions, and the Heavens," *Current Directions in Psychological Science* 18, no. 5 (2009): 264–68.

Chapter 11: Crafting Connection

182 **memories in her journal**: Roy F. Baumeister et al., "Thwarting the Need to Belong: Understanding the Interpersonal and Inner Effects of Social Exclusion," *Social and Personality Psychology Compass* 1, no. 1 (2007): 506–20.

184 **humans are relational beings**: David M. Sluss and Blake E. Ashforth, "Relational Identity and Identification: Defining Ourselves through Work Relationships," *Academy of Management Review* 32, no. 1 (2007): 9–32.

184 **100 to 150 people**: David M. Buss, "The Evolution of Happiness," *American Psychologist* 55, no. 1 (2000): 15; Glenn Geher et al., "You're Dead to Me! The Evolutionary Psychology of Social Estrangements and Social Transgressions," *Current Psychology* 40 (2021): 4516–30. Other researchers estimate group size between fifty and two hundred (Buss, "Evolution of Happiness" 15).

184 **core survival strategy**: P. Barchas, "A Sociophysiological Orientation to Small Groups," *Advances in Group Processes* 3 (1986): 209–46.

184 **by sharing labor**: Baumeister et al., "Thwarting the Need to Belong," 506–20.

184 **especially crave the care**: Roy F. Baumeister and Mark R. Leary, "The Need to Belong: Desire for Interpersonal Attachments as a Fundamental Human Motivation," *Psychological Bulletin* 117, no. 3 (1995): 497–529.

184 **cooperate**: Melvin Konner, "A Bold New Theory Proposes That Humans Tamed Themselves," *Atlantic*, March 2019, https://www.theatlantic.com/magazine/archive/2019/03/how-humans-tamed-themselves/580447.

184 **had to care**: Leonardo Christov-Moore et al., "Increasing Generosity by Disrupting Prefrontal Cortex," *Social Neuroscience* 12, no. 2 (2017): 174–81.

185 **two basic building blocks**: Edward L. Deci and Richard M. Ryan, "Self-Determination Research: Reflections and Future Directions," in *Handbook of Self-Determination Research*, ed. Edward L. Deci and Richard M. Ryan (New York: University of Rochester Press, 2002).

Notes

185 **belonging**: Baumeister and Leary, "Need to Belong," 497–529.

185 **help us work together**: Laura E. Stevens and Susan T. Fiske, "Motivation and Cognition in Social Life: A Social Survival Perspective," *Social Cognition* 13, no. 3 (1995): 189–214.

185 **form them easily**: Baumeister and Leary, "Need to Belong," 497–529.

185 **holiday card from a complete stranger**: Phillip R. Kunz and Michael Woolcott, "Season's Greetings: From My Status to Yours," *Social Science Research* 5, no. 3 (1976): 269–78.

185 **effortless in adversity**: Bibb Latané, Judith Eckman, and Virginia Joy, "Shared Stress and Interpersonal Attraction," *Journal of Experimental Social Psychology* 1 (1966): 80–94.

185 **make us feel seen, heard**: Sheldon Cohen and Thomas A. Wills, "Stress, Social Support, and the Buffering Hypothesis," *Psychological Bulletin* 98, no. 2 (1985): 310.

185 **relationship depth**: Baumeister and Leary, "Need to Belong," 497–529.

185 **same relatives**: Geher et al., "You're Dead to Me!," 4516–30.

185 **stand the test of time**: Baumeister and Leary, "Need to Belong," 497–529.

185 **trust**: Arlen C. Moller, Edward L. Deci, and Andrew J. Elliot, "Person-Level Relatedness and the Incremental Value of Relating," *Personality and Social Psychology Bulletin* 36, no. 6 (2010): 754–67.

185 **intimacy**: Kennon M. Sheldon and Jonathan C. Hilpert, "The Balanced Measure of Psychological Needs (BMPN) Scale: An Alternative Domain General Measure of Need Satisfaction," *Motivation and Emotion* 36 (2012): 439–51.

185 **truly care for us**: Sebastiano Costa, Nikos Ntoumanis, and Kimberley J. Bartholomew, "Predicting the Brighter and Darker Sides of Interpersonal Relationships: Does Psychological Need Thwarting Matter?," *Motivation and Emotion* 39 (2015): 11–24.

185 **especially under stress**: Ellen Skinner and Kathleen Edge, "Self-Determination, Coping, and Development," in *Handbook of Self-Determination Research*, ed. Edward L. Deci and Richard M. Ryan (Rochester, NY: University of Rochester Press, 2002), 297–337.

185 **we can rely on them**: Bram P. Buunk and Aukje Nauta, "Why Intraindividual Needs Are Not Enough: Human Motivation Is Primarily Social," *Psychological Inquiry* 11, no. 4 (2000): 279–83.

185 **our deepest relationships**: For a review, see Bram P. Buunk and Wilmar B. Schaufeli, "Reciprocity in Interpersonal Relationships: An Evolutionary Perspective on Its Importance for Health and Well-Being," *European Review of Social Psychology* 10, no. 1 (1999): DOI:10.1080/14792779943000080.

185 **function reciprocally**: Buunk and Nauta, "Why Intraindividual Needs Are Not Enough," 279–83.

185 **boosts our sense of self-worth**: Mark R. Leary et al., "Self-Esteem as an Interpersonal Monitor: The Sociometer Hypothesis," *Journal of Personality and Social Psychology* 68, no. 3 (1995): 518.

185 **and dignity**: John Wesley Kyle, *An Exploration of Human Dignity as a Foundation for Spiritual Leadership* (Derby, UK: University of Derby, 2020).

185 **rewires our brains**: Carolin Kieckhaefer, Leonhard Schilbach, and Danilo Bzdok, "Social Belonging: Brain Structure and Function Is Linked to Membership in Sports Teams, Religious Groups, and Social Clubs," *Cerebral Cortex* 33, no. 8 (2023): 4405–20.

185 **feel-good chemicals**: Anna J. Machin and Robin I. M. Dunbar, "The Brain Opioid Theory of Social Attachment: A Review of the Evidence," *Behaviour* 148, no. 9–10 (2011): 985–1025.

185 **exogenous opioids**: Jaak Panksepp, Stephen M. Siviy, and Lawrence A. Normansell, "Brain Opioids and Social Emotions," in *The Psychobiology of Attachment and Separation*, ed. Martin Reite and Tiffany Field (Orlando, FL: Academic Press, 1985), 3–49.

Notes

185 **even after factoring in**: James S. Goodwin et al., "The Effect of Marital Status on Stage, Treatment, and Survival of Cancer Patients," *Journal of the American Medical Association* 258, no. 21 (1987): 3125–30.

185 **protect against death**: Lynne C. Giles et al., "Effect of Social Networks on 10 Year Survival in Very Old Australians: The Australian Longitudinal Study of Aging," *Journal of Epidemiology & Community Health* 59, no. 7 (2005): 574–79.

185 **fifteen cigarettes**: Julianne Holt-Lunstad, Theodore F. Robles, and David A. Sbarra, "Advancing Social Connection as a Public Health Priority in the United States," *American Psychologist* 72, no. 6 (2017): 517.

185 **risks for dementia**: Ross Penninkilampi et al., "The Association between Social Engagement, Loneliness, and Risk of Dementia: A Systematic Review and Meta-Analysis," *Journal of Alzheimer's Disease* 66, no. 4 (2018): 1619–33.

185 **heart attack**: Baumeister and Leary, "Need to Belong," 497–529.

186 **mass shootings**: Jessica L. Lakin, Tanya L. Chartrand, and Robert M. Arkin, "I Am Too Just Like You: Nonconscious Mimicry as an Automatic Behavioral Response to Social Exclusion," *Psychological Science* 19, no. 8 (2008): 816–22.

186 **urgent public health concern**: Office of the US Surgeon General, "Our Epidemic of Loneliness and Isolation: The US Surgeon General's Advisory on the Healing Effects of Social Connection and Community," US Department of Health and Human Services, 2023, https://www.hhs.gov/sites/default/files/surgeon-general-social-connection-advisory.pdf.

186 **uproot our lives**: Geher et al., "You're Dead to Me!," 4516–30.

186 **extended kin networks**: Buss, "Evolution of Happiness," 15.

186 **sharply declining for decades**: Viji Diane Kannan and Peter J. Veazie, "US Trends in Social Isolation, Social Engagement, and Companionship—Nationally and by Age, Sex, Race/Ethnicity, Family Income, and Work Hours, 2003–2020," *Social Science & Medicine: Population Health* 21 (2023): 101331.

186 **in-person socializing has dropped**: Derek Thompson, "Why Americans Suddenly Stopped Hanging Out," *Atlantic*, February 14, 2024, https://www.theatlantic.com/ideas/archive/2024/02/america-decline-hanging-out/677451.

186 **fewer close friendships**: Buss, "Evolution of Happiness," 15.

186 **twenty fewer hours**: Kannan and Veazie, "US Trends in Social Isolation, Social Engagement, and Companionship," 101331.

186 **Demographic trends**: US Census Bureau, "Census Bureau Releases New Estimates on America's Families and Living Arrangements," press release CB22-TPS.99, November 17, 2022, https://www.census.gov/newsroom/press-releases/2022/americas-families-and-living-arrangements.html.

186 **39 percent of US adults**: Bridget Shovestul et al., "Risk Factors for Loneliness: The High Relative Importance of Age versus Other Factors," *PLOS One* 15, no. 2 (2020): e0229087.

186 **half experience persistent loneliness**: Shovestul et al., "Risk Factors for Loneliness," e0229087.

186 **twice as likely**: Robert D. Putnam, *The Upswing: How America Came Together a Century Ago and How We Can Do It Again* (New York: Simon and Schuster, 2020).

186 **reported having *zero* friends**: Jamie Ballard, "Millennials Are the Loneliest Generation," YouGov, July 30, 2019, https://today.yougov.com/society/articles/24577-loneliness-friendship-new-friends-poll-survey?redirect_from=%2Ftopics%2Flifestyle%2Farticles-reports%2F2019%2F07%2F30%2Floneliness-friendship-new-friends-poll-survey.

186 **As Derek Thompson writes**: Thompson, "Why Americans Suddenly Stopped Hanging Out."
186 **technology has played a key role**: Yalda T. Uhls, Nicole B. Ellison, and Kaveri Subrahmanyam, "Benefits and Costs of Social Media in Adolescence," *Pediatrics* 140, suppl. 2 (2017): S67–S70.
186 **doubled since 2015**: Emily A. Vogels, Risa Gelles-Watnick, and Navid Massarat, "Teens, Social Media and Technology 2022," Pew Research Center Report, August 10, 2022, https://www.pewresearch.org/internet/2022/08/10/teens-social-media-and-technology-2022.
186 **just one hour of daily screen time**: Jean M. Twenge and W. Keith Campbell, "Associations between Screen Time and Lower Psychological Well-Being among Children and Adolescents: Evidence from a Population-Based Study," *Preventive Medicine Reports* 12 (2018): 271–83.
186 **"a kind of ritual recession"**: Thompson, "Why Americans Suddenly Stopped Hanging Out."
187 **fewer than two in ten**: Thompson, "Why Americans Suddenly Stopped Hanging Out."
187 **near-historic lows**: Putnam, *Upswing*.
187 **shift to virtual and hybrid work**: Thompson, "Why Americans Suddenly Stopped Hanging Out."
187 **connection triggers can cut deep**: Sheldon and Hilpert, "Balanced Measure of Psychological Needs (BMPN) Scale," 439–51; Costa, Ntoumanis, and Bartholomew, "Predicting the Brighter and Darker Sides of Interpersonal Relationships," 11–24; Skinner and Edge, "Self-Determination, Coping, and Development"; Tiphaine Huyghebaert et al., "Leveraging Psychosocial Safety Climate to Prevent Ill-Being: The Mediating Role of Psychological Need Thwarting," *Journal of Vocational Behavior* 107 (2018): 111–25.
187 **rejection**: Costa, Ntoumanis, and Bartholomew, "Predicting the Brighter and Darker Sides of Interpersonal Relationships," 11–24.
187 **cause of anxiety**: Roy F. Baumeister and Dianne M. Tice, "Point-Counterpoints: Anxiety and Social Exclusion," *Journal of Social and Clinical Psychology* 9, no. 2 (1990): 165–95.
187 **and hostility**: Baumeister et al., "Thwarting the Need to Belong," 506–20.
187 **self-control, reasoning, and even IQ**: Roy F. Baumeister, Jean M. Twenge, and Christopher K. Nuss, "Effects of Social Exclusion on Cognitive Processes: Anticipated Aloneness Reduces Intelligent Thought," *Journal of Personality and Social Psychology* 83, no. 4 (2002): 817.
187 **respond to rejection**: Kipling D. Williams, "Ostracism: A Temporal Need-Threat Model," *Advances in Experimental Social Psychology* 41 (2009): 275–314.
187 **Studies show**: "Ostracism," 275–314.
187 **neglect**: Nicolas Gillet et al., "The Effects of Organizational Factors, Psychological Need Satisfaction and Thwarting, and Affective Commitment on Workers' Well-Being and Turnover Intentions," *Le travail humain* 78, no. 2 (2015): 119–40.
187 **fail to show us care**: Meredith Rocchi et al., "Assessing Need-Supportive and Need-Thwarting Interpersonal Behaviours: The Interpersonal Behaviours Questionnaire (IBQ)," *Personality and Individual Differences* 104 (2017): 423–33; Filipe Rodrigues et al., "The Role of Dark-Side of Motivation and Intention to Continue in Exercise: A Self-Determination Theory Approach," *Scandinavian Journal of Psychology* 60, no. 6 (2019): 585–95.
187 **or concern**: Tiphaine Huyghebaert-Zouaghi et al., "Advancing the Conceptualization and Measurement of Psychological Need States: A 3×3 Model Based on Self-Determination Theory," *Journal of Career Assessment* 29, no. 3 (2021): 396–421.
187 **cold**: Rodrigues et al., "Role of Dark-Side of Motivation and Intention to Continue in Exercise," 585–95.

Notes

187 **distant interactions**: Rocchi et al., "Assessing Need-Supportive and Need-Thwarting Interpersonal Behaviours," 423–33.

187 **preferential treatment**: Buss, "Evolution of Happiness," 15.

187 **when we behave the way they want**: Maarten Vansteenkiste, Christopher P. Niemiec, and Bart Soenens, "The Development of the Five Mini-Theories of Self-Determination Theory: An Historical Overview, Emerging Trends, and Future Directions," in *The Decade Ahead: Theoretical Perspectives on Motivation and Achievement*, ed. Timothy C. Urdan and Stuart A. Karabenick (Bingley, UK: Emerald, 2010), 105–65.

188 **harmful intentions**: Ellen Giebels and Onne Janssen, "Conflict Stress and Reduced Well-Being at Work: The Buffering Effect of Third-Party Help," in *Conflict in Organizations: Beyond Effectiveness and Performance*, ed. Fred Zijlstra (New York: Psychology Press, 2020), 137–55.

188 **personal interests, goals**: James A. Wall Jr. and Ronda Roberts Callister, "Conflict and Its Management," *Journal of Management* 21, no. 3 (1995): 515–58.

188 **or status**: Buss, "Evolution of Happiness," 15.

188 **challenges our core desire**: Giebels and Janssen, "Conflict Stress and Reduced Well-Being at Work," 137–55.

188 **dehumanizing or hurtful behavior**: Lei Cheng et al., "Objectification Limits Authenticity: Exploring the Relations between Objectification, Perceived Authenticity, and Subjective Well-Being," *British Journal of Social Psychology* 61, no. 2 (2022): 622–43.

188 **abuse, attacks, punishment**: Vello Hein, Andre Koka, and Martin S. Hagger, "Relationships between Perceived Teachers' Controlling Behaviour, Psychological Need Thwarting, Anger and Bullying Behaviour in High-School Students," *Journal of Adolescence* 42 (2015): 103–14.

188 **manipulation**: Majid Murad et al., "The Influence of Despotic Leadership on Counterproductive Work Behavior among Police Personnel: Role of Emotional Exhaustion and Organizational Cynicism," *Journal of Police and Criminal Psychology* 36, no. 3 (2021): 603–15.

188 **humiliation**: Hein, Koka, and Hagger, "Relationships between Perceived Teachers' Controlling Behaviour, Psychological Need Thwarting, Anger and Bullying Behaviour in High-School Students," 103–14.

188 **being exploited**: Cheng et al., "Objectification Limits Authenticity," 622–43.

188 **overtly**: Ellen Skinner, Sandy Johnson, and Tatiana Snyder, "Six Dimensions of Parenting: A Motivational Model," *Parenting: Science and Practice* 5, no. 2 (2005): 175–235; Nikita Bhavsar et al., "Conceptualizing and Testing a New Tripartite Measure of Coach Interpersonal Behaviors," *Psychology of Sport and Exercise* 44 (2019): 107–20.

188 **deep feelings of shame**: John Wesley Kyle, *An Exploration of Human Dignity as a Foundation for Spiritual Leadership* (Derby, UK: University of Derby, 2020).

188 **84 percent of people**: Darren C. Treadway et al., "Political Skill and the Job Performance of Bullies," *Journal of Managerial Psychology* 28, no. 3 (2013): 273–89.

188 **socially connected superstars**: Iain Coyne, Jane Craig, and Penelope Smith-Lee Chong, "Workplace Bullying in a Group Context," *British Journal of Guidance & Counselling* 32, no. 3 (2004): 301–17.

188 **especially women**: Helge Hoel, Cary L. Cooper, and Brian Faragher, "The Experience of Bullying in Great Britain: The Impact of Organizational Status," *European Journal of Work and Organizational Psychology* 10, no. 4 (2001): 443–65.

Notes

188 **betrayal**: Buss, "Evolution of Happiness," 15.

188 **someone we feel is "safe"**: Geher et al., "You're Dead to Me!," 4516–30.

188 **violates our expectations**: Bradfield, Murray, and Karl Aquino, "The effects of blame attributions and offender likableness on forgiveness and revenge in the workplace," *Journal of management* 25, no. 5 (1999): 607–31.

188 **ancestors avoid exploitation**: Geher et al., "You're Dead to Me!," 4516–30.

188 **disproportionately triggering**: J. C. Badcock et al., "Position Statement: Addressing Social Isolation and Loneliness and the Power of Human Connection," *Global Initiative on Loneliness and Connection (GILC)* (2022): 1–43.

188 **eleven million pieces**: Tor Norretranders, *The User Illusion: Cutting Consciousness Down to Size* (New York: Penguin, 1999).

189 **over detect interpersonal threat**: Hyesung Grace Hwang, "Development of Social Exclusion Detection: Behavioral and Physiological Correlates" (PhD diss., Washington University in St. Louis, 2018).

189 **notice a mate's infidelity**: Buss, "Evolution of Happiness," 15.

189 **no actual betrayal**: Ayala Pines and Elliot Aronson, "Antecedents, Correlates, and Consequences of Sexual Jealousy," *Journal of Personality* 51, no. 1 (1983): 108–36.

189 **spite**: Yaniv Kanat-Maymon et al., "The Role of Basic Need Fulfillment in Academic Dishonesty: A Self-Determination Theory Perspective," *Contemporary Educational Psychology* 43 (2015): 1–9; Julien S. Bureau et al., "Investigating How Autonomy-Supportive Teaching Moderates the Relation between Student Honesty and Premeditated Cheating," *British Journal of Educational Psychology* 92, no. 1 (2022): 175–93.

189 **aggression**: Jean M. Twenge et al., "If You Can't Join Them, Beat Them: Effects of Social Exclusion on Aggressive Behavior," *Journal of Personality and Social Psychology* 81, no. 6 (2001): 1058.

189 **popularity**: Richard M. Ryan et al., "The American Dream in Russia: Extrinsic Aspirations and Well-Being in Two Cultures," *Personality and Social Psychology Bulletin* 25, no. 12 (1999): 1509–24.

189 **validation**: Alain Van Hiel and Maarten Vansteenkiste, "Ambitions Fulfilled? The Effects of Intrinsic and Extrinsic Goal Attainment on Older Adults' Ego-Integrity and Death Attitudes," *International Journal of Aging and Human Development* 68, no. 1 (2009): 27–51.

189 **seclusion**: Kennon M. Sheldon and Alexander Gunz, "Psychological Needs as Basic Motives, Not Just Experiential Requirements," *Journal of Personality* 77, no. 5 (2009): 1467–92.

189 **pretending**: Daniel Freeman et al., "Psychological Investigation of the Structure of Paranoia in a Non-Clinical Population," *British Journal of Psychiatry* 186, no. 5 (2005): 427–35.

189 **biggest impact on our mental health and well-being**: Specifically, warding off depression. Baumeister and Leary, "Need to Belong," 57–89.

189 **yielding diminishing returns**: Linda S. Ruehlman and Sharlene A. Wolchik, "Personal Goals and Interpersonal Support and Hindrance as Factors in Psychological Distress and Well-Being," *Journal of Personality and Social Psychology* 55, no. 2 (1988): 293.

189 **often romantic:** James S. Goodwin et al., "The Effect of Marital Status on Stage, Treatment, and Survival of Cancer Patients," *Journal of the American Medical Association* 258, no. 21 (1987): 3125–30.

189 **but they don't have to be**: Machin and Dunbar, "Brain Opioid Theory of Social Attachment," 985–1025.

190 **can be virtually indistinguishable**: Buss, "Evolution of Happiness," 15.

Notes

195 **a surge of red-hot rage:** See Daniel Sznycer, Aaron Sell, and Alexandre Dumont, "How Anger Works," *Evolution and Human Behavior* 43, no. 2 (2022): 122–32.

195 **is other people**: Todd B. Kashdan et al., "What Triggers Anger in Everyday Life? Links to the Intensity, Control, and Regulation of these Emotions, and Personality Traits," *Journal of Personality* 84, no. 6 (2016): 73–49.

195 **record or the wrong fall flat:** Sznycer, Sell, and Dumont, "How Anger Works," 122–32.

195 **neither good nor bad:** K. J. Nimisha, "Analysis of Triggers, Expression and Management of Anger: Cross Sectional survey" (PhD diss., JKK Nattraja College of Pharmacy, Kumarapalayam, 2021).

195 **stand up for ourselves**: Aaron N. Sell and Anthony C. Lopez, "Emotional Underpinnings of War: An Evolutionary Analysis of Anger and Hatred," in *The Handbook of Collective Violence: Current Developments and Understanding*, ed. Carol A. Ireland et al. (Abingdon, UK: Routledge, 2020), 31–46; Agneta H. Fischer and Ira J. Roseman, "Beat Them or Ban Them: The Characteristics and Social Functions of Anger and Contempt," *Journal of Personality and Social Psychology* 93, no. 1 (2007): 103.

195 **aggression or spite**: Sznycer, Sell, and Dumont, "How Anger Works," 122–32.

195 **offense rumination**: Nathaniel G. Wade et al., "Measuring State-Specific Rumination: Development of the Rumination about an Interpersonal Offense Scale," *Journal of Counseling Psychology* 55, no. 3 (2008): 419.

195 **elaborate revenge fantasies**: MeowLan Evelyn Chan and Daniel J. McAllister, "Abusive Supervision through the Lens of Employee State Paranoia," *Academy of Management Review* 39, no. 1 (2014): 44–66.

195 **hampers conflict resolution**: K. Lira Yoon and Jutta Joormann, "Is Timing Everything? Sequential Effects of Rumination and Distraction on Interpersonal Problem Solving," *Cognitive Therapy and Research* 36 (2012): 165–72.

195 **breeds paranoia**: Roderick M. Kramer, "The Sinister Attribution Error: Paranoid Cognition and Collective Distrust in Organizations," *Motivation and Emotion* 18 (1994): 199–230.

195 **and anger**: Aslı Ascıoglu Onal and İlhan Yalcın, "Forgiveness of Others and Self-Forgiveness: The Predictive Role of Cognitive Distortions, Empathy, and Rumination," *Eurasian Journal of Educational Research* 17, no. 68 (2017): 97–120.

195 **observation about human nature**: Cavan Sieczkowski, "Former CIA Officer: Listen to Your Enemy Because 'Everybody Believes They Are the Good Guy,'" *Huffpost*, June 14, 2016, https://www.huffpost.com/entry/amaryllis-fox-undercover-cia-video_n_57600d31e4b0e4fe5143afc6.

195 **self-importance and entitlement skyrocket**: Emily Maria Zitek, *Feeling Wronged Leads to Entitlement and Selfishness* (Stanford, CA: Stanford University, 2010).

196 **demonize those who've wronged us**: Dennis R. Combs et al., "The Ambiguous Intentions Hostility Questionnaire (AIHQ): A New Measure for Evaluating Hostile Social-Cognitive Biases in Paranoia," *Cognitive Neuropsychiatry* 12, no. 2 (2007): 128–43.

196 **bad guys bias**: This shadow habit goes by many names in psychological research, including hypermentalizing, projective identification, sinister attribution error, hostile social cognitive bias, and interpretation bias (Alessandro Grecucci et al., "Anxious Ultimatums: How Anxiety Disorders Affect Socioeconomic Behaviour," *Cognition & Emotion* 27, no. 2 (2013): 230–44; Roderic, M. Kramer, "The Sinister Attribution Error: Paranoid Cognition and Collective Distrust in Organizations," *Motivation and Emotion* 18 (1994): 199–230; Combs et al., "Ambiguous Intentions Hostility Questionnaire (AIHQ)," 128–43; Paula Hertel et al.,

"Looking on the Dark Side: Rumination and Cognitive-Bias Modification," *Clinical Psychological Science* 2, no. 6 (2014): 714–26.

196 **sneaky habit fueling**: Hertel et al., "Looking on the Dark Side," 714–26.

196 **precious mental resources**: Kramer, "Sinister Attribution Error," 199–230.

196 **aggressive**: Paul J. C. Adachi and Teena Willoughby, "The Effect of Video Game Competition and Violence on Aggressive Behavior: Which Characteristic Has the Greatest Influence?," *Psychology of Violence* 1, no. 4 (2011): 259.

196 **socially inappropriate behavior**: Grecucci et al., "Anxious Ultimatums," 230–44.

196 **intensifies our feuds**: Mark Snyder and William B. Swann Jr., "Behavioral Confirmation in Social Interaction: From Social Perception to Social Reality," *Journal of Experimental Social Psychology* 14, no. 2 (1978): 148–62.

196 **the brain's exploration network**: Michael Platt et al., "Perspective Taking: A Brain Hack That Can Help You Make Better Decisions," *Knowledge at Wharton*, March 22, 2021, https://knowledge.wharton.upenn.edu/article/perspective-taking-brain-hack-can-help-make-better-decisions.

197 **creative perspective taking**: Xiaofei Wu et al., "Role of Creativity in the Effectiveness of Cognitive Reappraisal," *Frontiers in Psychology* 8 (2017): 269718.

199 **preserving the sacred**: Dirk Van Dierendonck, "Spirituality as an Essential Determinant for the Good Life, Its Importance Relative to Self-Determinant Psychological Needs," *Journal of Happiness Studies* 13 (2012): 685–700.

199 **deep, authentic pursuit of the sacred**: Sam A. Hardy et al., "Adolescent Religious Motivation: A Self-Determination Theory Approach," *International Journal for the Psychology of Religion* 32, no. 1 (2022): 16–30; T. T. Kneezel and R. M. Ryan, "Not Just a Reflection of Parents: God as a Source of Support and Nurturance," poster presented at the second International Conference on Self-Determination Theory, Ottawa, Ontario, 2004.

199 **lies awe:** Peter O. Kearns and James M. Tyler, "Examining the Relationship between Awe, Spirituality, and Religiosity," *Psychology of Religion and Spirituality* 14, no. 4 (2022): 436.

199 **greater than ourselves**: Susan K. Chen and Myriam Mongrain, "Awe and the Interconnected Self," *Journal of Positive Psychology* 16, no. 6 (2021): 770–78.

202 **Offense rumination**: Edward L. Deci and Richard M. Ryan, "The 'What' and 'Why' of Goal Pursuits: Human Needs and the Self-Determination of Behavior," *Psychological Inquiry* 11, no. 4 (2000): 227–68.

202 **Paranoia**: Freeman et al., "Psychological Investigation of the Structure of Paranoia in a Non-Clinical Population," 427–35.

202 **hurting others to feel better**: Richard M. Ryan et al., "Building a Science of Motivated Persons: Self-Determination Theory's Empirical Approach to Human Experience and the Regulation of Behavior," *Motivation Science* 7, no. 2 (2021): 97; Henry A. Murray, *Explorations in Personality* (New York: Oxford University Press, 1938).

202 **those who've wronged us:** Jaesub Lee, Jingpei J. C. Lim, and Robert L. Heath, "Coping with Workplace Bullying through NAVER: Effects of LMX Relational Concerns and Cultural Differences," *International Journal of Business Communication* 58, no. 1 (2021): 79–105; Murray, *Explorations in Personality*.

202 **Hostility**: Baumeister et al., "Thwarting the Need to Belong," 506–20.

202 **status symbols**: Alain Van Hiel and Maarten Vansteenkiste, "Ambitions Fulfilled? The Effects of Intrinsic and Extrinsic Goal Attainment on Older Adults' Ego-Integrity and Death Attitudes," *International Journal of Aging and Human Development* 68, no. 1 (2009): 27–51; Joke Verstuyf

et al., "Motivational Dynamics of Eating Regulation: A Self-Determination Theory Perspective," *International Journal of Behavioral Nutrition and Physical Activity* 9, no. 1 (2012): 1–16.

202 **focused on fame**: Tim Kasser and Richard M. Ryan, "Further Examining the American Dream: Differential Correlates of Intrinsic and Extrinsic Goals," *Personality and Social Psychology Bulletin* 22, no. 3 (1996): 280–87.

202 **needing social recognition**: Van Hiel and Vansteenkiste, "Ambitions Fulfilled?," 27–51; Valerie Good et al., "A Self-Determination Theory-Based Meta-Analysis on the Differential Effects of Intrinsic and Extrinsic Motivation on Salesperson Performance," *Journal of the Academy of Marketing Science* 50, no. 3 (2022): 586–614.

202 **praise for our contributions**: Good et al., "Self-Determination Theory-Based Meta-Analysis on the Differential Effects of Intrinsic and Extrinsic Motivation on Salesperson Performance," 586–614.

202 **flattery and compliments**: Freeman et al., "Psychological Investigation of the Structure of Paranoia in a Non-Clinical Population," 427–35.

202 **Shying away from confrontation**: Nicolas Bacon and Paul Blyton, "Conflict for Mutual Gains?," *Journal of Management Studies* 44, no. 5 (2007): 814–34; Robert J. Bies et al., "Beyond Distrust, Beyond Distrust: "Getting Even" and the Need for Revenge" in *Trust in Organizations: Frontiers of Theory and Research*, ed. Roderick M. Kramer and Tom R. Tyler (Thousand Oaks, CA: Sage, 1996), 246–60.

202 **doing nothing**: Lee, Lim, and Heath, "Coping with Workplace Bullying through NAVER," 79–105.

202 **staying silent**: Qin Xu et al., "Abusive Supervision, High-Performance Work Systems, and Subordinate Silence," *Personnel Review* 49, no. 8 (2020): 1637–53; Wen Wu et al., "Needs Frustration Makes Me Silent: Workplace Ostracism and Newcomers' Voice Behavior," *Journal of Management & Organization* 25, no. 5 (2019): 635–52; Williams, "Ostracism," 275–314.

202 **Conforming**: Freeman et al., "Psychological Investigation of the Structure of Paranoia in a Non-Clinical Population," 427–35; Kipling D. Williams, Christopher K. T. Cheung, and Wilma Choi, "Cyberostracism: Effects of Being Ignored over the Internet," *Journal of Personality and Social Psychology* 79, no. 5 (2000): 748.

202 **or expectations**: Jessica L. Lakin, Tanya L. Chartrand, and Robert M. Arkin, "I Am Too Just Like You: Nonconscious Mimicry as an Automatic Behavioral Response to Social Exclusion," *Psychological Science* 19, no. 8 (2008): 816–22.

202 **unwanted requests**: Williams, "Ostracism," 275–314.

202 **Withdrawing**: Roy F. Baumeister et al., "Social Exclusion Impairs Self-Regulation," *Journal of Personality and Social Psychology* 88, no. 4 (2005): 589.

202 **or detaching**: Sheldon and Gunz, "Psychological Needs as Basic Motives," 1467–92.

202 **to protect against further hurt**: Freeman et al., "Psychological Investigation of the Structure of Paranoia in a Non-Clinical Population," 427–35; Sheldon and Gunz, "Psychological Needs as Basic Motives," 1467–92.

202 **Choosing solitary activities**: Sheldon and Gunz, "Psychological Needs as Basic Motives," 1467–92.

202 **people are "dead to us"**: Baumeister et al., "Social Exclusion Impairs Self-Regulation," 589.

203 **envy of the world**: Ross Jones and Ramon Rosario, "'There's Nothing Left': After Decades of Decline, Highland Park Fights for a Future," WXYZ Detroit, September 22, 2022, https://

www.wxyz.com/news/local-news/investigations/theres-nothing-left-after-decades-of-decline-highland-park-fights-for-a-future.

203 **the auto industry had fled**: Ford, "Highland Park," accessed September 12, 2024, https://corporate.ford.com/articles/history/highland-park.html; Jones and Rosario, "There's Nothing Left."

203 **fell into decline**: Meg Dunn, "Turning Pain into Power: How a Grieving Mother Transformed a Neglected Block Near Detroit into a Village of Beauty and Opportunity," CNN, June 16, 2023, https://www.cnn.com/2023/06/16/us/detroit-real-estate-safety-education-sustainability-opportunity-grief-cnnheroes/index.html; Anderson Cooper, "Mama Shu: Turning Loss into Love," *All There Is with Anderson Cooper*, January 24, 2024, https://www.cnn.com/audio/podcasts/all-there-is-with-anderson-cooper/episodes/c74c5578-741e-11ee-b574-67f673692459.

203 **under a neighbor's supervision**: Zenobia Jeffries Warfield, "It Was a Blighted City Block. But This Woman Is Turning It into a Solar-Powered Ecovillage," Yes!, December 7, 2016, https://www.yesmagazine.org/democracy/2016/12/07/it-was-a-blighted-city-block-but-this-woman-is-turning-it-into-a-solar-powered-ecovillage.

204 **removing him from life support**: Dunn, "Turning Pain into Power"; Cooper, "Mama Shu."

204 **Every day on her way to work**: Erin Rose, "Life without Money in Detroit's Survival Economy," PositiveDetroit, January 17, 2017, http://www.positivedetroit.net/2017/01.

204 **she pictured how beautiful**: Dunn, "Turning Pain into Power"; Cooper, "Mama Shu."

204 **mend her own**: Shamayim Harris, "From Blight to Beauty," TEDxDetroit, TedX Talks, posted on YouTube, February 27, 2017, https://www.youtube.com/watch?v=cGdJyyolmSw.

204 **loan from her sister**: Luke Gannon and Reggie Rucker, "It Takes an Avalon Village," Institute for Local Self-Reliance, March 21, 2024, https://ilsr.org/it-takes-an-avalon-village.

204 **she offered $3,000**: Dunn, "Turning Pain into Power"; Cooper, "Mama Shu."

204 **for its children**: Eleanore Catolico, "A Mother Mourns Again, Her Beloved Community Mourns with Her," Bridge Detroit, February 8, 2021, https://www.bridgedetroit.com/a-mother-mourns-again-her-beloved-community-mourns-with-her.

204 **below the poverty line**: Dunn, "Turning Pain into Power"; Cooper, "Mama Shu."

204 **rising like a phoenix**: Cooper, "Mama Shu."

205 **would give children a safe place**: Catolico, "Mother Mourns Again."

205 **lost family members to violence**: Dunn, "Turning Pain into Power"; Cooper, "Mama Shu."

205 **revitalizing Avalon Street**: Dunn, "Turning Pain into Power"; Cooper, "Mama Shu."

205 **at point-blank range**: Karen Drew, "Highland Park's Mama Shu Fights for Justice after Son Murdered While Protecting Neighborhood," Click on Detroit, March 10, 2022, https://www.clickondetroit.com/news/defenders/2022/03/10/highland-parks-mama-shu-fights-for-justice-after-son-murdered-while-protecting-neighborhood.

205 **struck by lightning twice**: Charlie LeDuff, "Two Sons Lost to Violence, 'Mama Shu' Vows to Carry on in Highland Park," Deadline Detroit, February 4, 2021, https://deadlinedetroit.com/articles/27278/leduff_two_sons_lost_to_violence_mama_shu_vows_to_carry_on_in_highland_park.

205 **considered giving up**: Nushrat Rahman, "Highland Park's 'Mama Shu' Is Among USA Today's 2024 Women of the Year," *Detroit Free Press*, February 29, 2024. https://www.freep.com/story/news/local/michigan/2024/02/29/shamayim-mama-shu-harris-highland-park-michigan-women-of-year/72094887007/.

Notes

205 **mobilizing a security team**: Gannon and Rucker, "It Takes an Avalon Village."
205 **erecting fences**: Slone Terranella, "Highland Park Leader's Son Fatally Shot: 'The Pain Is Indescribable,'" *Detroit Free Press*, February 1, 2021, https://www.freep.com/story/news/local/michigan/wayne/2021/02/01/highland-park-mama-shu-son-chinyelu-humphrey/4333714001.
205 **spans three blocks**: Dunn, "Turning Pain into Power"; Cooper, "Mama Shu."
205 **organic produce to the community**: Nushrat Rahman, "Treasured Community Garden in Highland Park Could Be Uprooted in Land Dispute," Bridge Detroit, September 2, 2023, https://www.bridgedetroit.com/treasured-community-garden-in-highland-park-could-be-uprooted-in-land-dispute.
205 **Goddess Marketplace**: CW50Detroit, "Highland Park Mom Transforms Blighted Neighborhood to Beauty While Suffering Devastating Loss," March 21, 2022, https://www.cbsnews.com/detroit/news/highland-park-mom-transforms-blighted-neighborhood-to-beauty-all-while-suffering-devastating-loss.
206 **STEAM lab**: Dunn, "Turning Pain into Power"; Cooper, "Mama Shu."
206 **wine-tasting room**: Gannon and Rucker, "It Takes an Avalon Village."
206 **music studio**: Dunn, "Turning Pain into Power"; Cooper, "Mama Shu."
206 **Healing House offers**: Dian Zhang, "Avalon Village: A Blighted Detroit Neighborhood Turned Eco-Village," *Architect*, May 28, 2016, https://www.architectmagazine.com/design/avalon-village-a-blighted-detroit-neighborhood-turned-eco-village_o.
206 **solar streetlights**: Dunn, "Turning Pain into Power"; Cooper, "Mama Shu."
206 **peace and solace**: Avalon Village official website, "About Avalon Village," accessed September 12, 2024, https://www.theavalonvillage.org/about-the-village.
206 **market-rate housing**: Dunn, "Turning Pain into Power"; Cooper, "Mama Shu."
206 **"grief into glory"**: Harris, "From Blight to Beauty."
206 **forever carry the pain**: Cooper, "Mama Shu."
206 **"break your spirit"**: Harris, "From Blight to Beauty."
206 **end up being a choice**: Cooper, "Mama Shu."
207 **positive and powerful tendencies**: Maarten Vansteenkiste and Richard M. Ryan, "On Psychological Growth And Vulnerability: Basic Psychological Need Satisfaction and Need Frustration as a Unifying Principle," *Journal of Psychotherapy Integration* 23, no. 3 (2013): 263.
207 **energy, joy, and well-being**: Edward L. Deci and Richard M. Ryan, "Motivation, Personality, and Development within Embedded Social Contexts: An Overview of Self-Determination Theory," in *The Oxford Handbook of Human Motivation*, ed. Richar M. Ryan (Oxford: Oxford University Press, 2012), 85–107.
207 **stronger performance**: Christopher P. Cerasoli, Jessica M. Nicklin, and Alexander S. Nassrelgrgawi, "Performance, Incentives, and Needs for Autonomy, Competence, and Relatedness: A Meta-Analysis," *Motivation and Emotion* 40 (2016): 781–813.
207 **deeper learning**: Maarten Vansteenkiste, Christopher P. Niemiec, and Bart Soenens, "The Development of the Five Mini-Theories of Self-Determination Theory: An Historical Overview, Emerging Trends, and Future Directions," in *The Decade Ahead: Theoretical Perspectives on Motivation and Achievement*, ed. Timothy C. Urdan and Stuart A. Karabenick (Bingley, UK: Emerald, 2010), 105–65.
207 **charting a life**: Richard M. Ryan, Veronika Huta, and Edward L. Deci, "Living Well: A Self-Determination Theory Perspective on Eudaimonia," *Journal of Happiness Studies* 9 (2008): 139–70.

207 **purpose, peace**: Michaéla C. Schippers and Niklas Ziegler, "Life Crafting as a Way to Find Purpose and Meaning in Life," *Frontiers in Psychology* 10 (2019): 2778; Netta Weinstein and Holley S. Hodgins, "The Moderating Role of Autonomy and Control on the Benefits of Written Emotion Expression," *Personality and Social Psychology Bulletin* 35, no. 3 (2009): 351–64.

207 **physical health**: Richard M. Ryan et al., "We Know This Much Is (Meta-Analytically) True: A Meta-Review of Meta-Analytic Findings Evaluating Self-Determination Theory," *Psychological Bulletin* 148, no. 11–12 (2022): 813.

207 **longevity**: Vansteenkiste, Niemiec, and Soenens, "Development of the Five Mini-Theories of Self-Determination Theory," 105–65.

208 **better parent**: Vansteenkiste, Niemiec, and Soenens, "Development of the Five Mini-Theories of Self-Determination Theory," 105–65.

208 **partner**: Richard M. Ryan et al., "Building a Science of Motivated Persons: Self-Determination Theory's Empirical Approach to Human Experience and the Regulation of Behavior," *Motivation Science* 7, no. 2 (2021): 97.

208 **friend**: On better social functioning, social competence, empathy, secure attachments, see Vansteenkiste, Niemiec, and Soenens, "Development of the Five Mini-Theories of Self-Determination Theory," 140.

208 **colleague**: Ryan et al., "We Know This Much Is (Meta-Analytically) True," 813; Jean Fox Craig, "Employee Empowerment, Self-Determination Theory, and Employee Engagement: A Mediation Model" (PhD diss., Oklahoma University, 2017).

208 **and citizen**: Deci and Ryan, "Motivation, Personality, and Development within Embedded Social Contexts," 85–107.

208 **generous with others**: Richard M. Ryan, Bart Soenens, and Maarten Vansteenkiste, "Reflections on Self-Determination Theory as an Organizing Framework for Personality Psychology: Interfaces, Integrations, Issues, and Unfinished Business," *Journal of Personality* 87, no. 1 (2019): 115–45.

209 *twice as effective*: Christopher J. Boyce, Alex M. Wood, and Nattavudh Powdthavee, "Is Personality Fixed? Personality Changes as Much as "Variable" Economic Factors and More Strongly Predicts Changes to Life Satisfaction," *Social Indicators Research* 111 (2013): 287–305.

210 **"the dictator within"**: Steven C. Hayes, *A Liberated Mind: How to Pivot toward What Matters* (New York: Penguin, 2020).

211 **Archetypes of growth and transformation**: Francisco Huanac, "Sacred Spiral: Meaning of the Ancient Symbol of the Goddess," *Spells8 Blog*, February 21, 2022, https://spells8.com/sacred-spiral-meaning.

Epilogue

215 **the plasma of ten thousand people**: Immune Deficiency Foundation, "Immunoglobulin Replacement Therapy: What It Is and Why Donating Plasma Matters," October 8, 2018, https://primaryimmune.org/resources/news-articles/what-ig-replacement-therapy-and-why-donating-plasma-matters.

Appendix B

226 **Chronic stress**: Belinda Bruwer et al., "Psychometric Properties of the Multidimensional Scale of Perceived Social Support in Youth," *Comprehensive Psychiatry* 49, no. 2 (2008):

195–201; Katie A. McLaughlin et al., "Childhood Adversity, Adult Stressful Life Events, and Risk of Past-Year Psychiatric Disorder: A Test of the Stress Sensitization Hypothesis in a Population-Based Sample of Adults," *Psychological Medicine* 40, no. 10 (2010): 1647–58; Paula S. Nurius, Edwina Uehara, and Douglas F. Zatzick, "Intersection of Stress, Social Disadvantage, and Life Course Processes: Reframing Trauma and Mental Health," *American Journal of Psychiatric Rehabilitation* 16, no. 2 (2013): 91–114. On stressful life experiences in the past year, see Sharain Suliman et al., "Cumulative Effect of Multiple Trauma on Symptoms of Posttraumatic Stress Disorder, Anxiety, and Depression in Adolescents," *Comprehensive Psychiatry* 50, no. 2 (2009): 121–27. On current life stress, see George A. Bonanno et al., "What Predicts Psychological Resilience after Disaster? The Role of Demographics, Resources, and Life Stress," *Journal of Consulting and Clinical Psychology* 75, no. 5 (2007): 671. On even minor chronic stress, see Scott M. Monroe and Kate L. Harkness, "Life Stress, the 'Kindling' Hypothesis, and the Recurrence of Depression: Considerations from a Life Stress Perspective," *Psychological Review* 112, no. 2 (2005): 417.

226 **Health concerns**: On colds, flu, and fatigue, see Douglas Paton, Leigh Smith, and John Violanti, "Disaster Response: Risk, Vulnerability and Resilience," *Disaster Prevention and Management: An International Journal* 9, no. 3 (2000): 173–80.

226 **job or career stress**: Paton, Smith, and Violanti, "Disaster Response," 173–80.

226 **job change**: Changes in work or job. McLaughlin et al., "Childhood Adversity, Adult Stressful Life Events, and Risk of Past-Year Psychiatric Disorder," 1647–58.

226 **Mental health challenges**: Carmen Valiente et al., "A Symptom-Based Definition of Resilience in Times of Pandemics: Patterns of Psychological Responses over Time and Their Predictors," *European Journal of Psychotraumatology* 12, no. 1 (2021): 1871555; Paula P. Schnurr, Matthew J. Friedman, and Stanley D. Rosenberg, "Premilitary MMPI Scores as Predictors of Combat-Related PTSD Symptoms," *American Journal of Psychiatry* 150, no. 3 (1993): 479–83; Chris R. Brewin, Bernice Andrews, and John D. Valentine, "Meta-Analysis of Risk Factors for Posttraumatic Stress Disorder in Trauma-Exposed Adults," *Journal of Consulting and Clinical Psychology* 68, no. 5 (2000): 748.

226 **Problems with friends, colleagues, or family**: McLaughlin et al., "Childhood Adversity, Adult Stressful Life Events, and Risk of Past-Year Psychiatric Disorder," 1647–58; Elizabeth A. Newnham et al., "Youth Mental Health after Civil War: The Importance of Daily Stressors," *British Journal of Psychiatry* 206, no. 2 (2015): 116–21; Miller et al., "Daily Stressors, War Experiences, and Mental Health in Afghanistan," 611–38.

226 **Marital or romantic challenges**: See Constance Hammen, "Stress and Depression," *Annual Review of Clinical Psychology* 1 (2005): 293–319.

226 **Loved ones experiencing**: McLaughlin et al., "Childhood Adversity, Adult Stressful Life Events, and Risk of Past-Year Psychiatric Disorder," 1647–58.

227 **Financial problems**: Miller et al., "Daily Stressors, War Experiences, and Mental Health in Afghanistan," 611–38. On poverty, see Hammen, "Stress and Depression," 293–319. On economic hardship, see John W. Lynch, George A. Kaplan, and Sarah J. Shema, "Cumulative Impact of Sustained Economic Hardship on Physical, Cognitive, Psychological, and Social Functioning," *New England Journal of Medicine* 337, no. 26 (1997): 1889–95. On major financial crisis, see Cristina A. Fernandez et al., "Assessing the Relationship between Psychosocial Stressors and Psychiatric Resilience among Chilean Disaster Survivors," *British Journal of Psychiatry* 217, no. 5 (2020): 630–37.

Notes

227 **Legal trouble**: McLaughlin et al., "Childhood Adversity, Adult Stressful Life Events, and Risk of Past-Year Psychiatric Disorder," 1647–58.

227 **serious conflict**: Fernandez et al., "Assessing the Relationship between Psychosocial Stressors and Psychiatric Resilience among Chilean Disaster Survivors," 630–37.

227 **Marital separation or divorce**: Hammen, "Stress and Depression," 293–319. On marital or relationship separation, see Fernandez et al., "Assessing the Relationship between Psychosocial Stressors and Psychiatric Resilience among Chilean Disaster Survivors," 630–37.

227 **illness**: Fernandez et al., "Assessing the Relationship between Psychosocial Stressors and Psychiatric Resilience among Chilean Disaster Survivors," 630–37.

227 **injury**: On health concerns, see Miller et al., "Daily Stressors, War Experiences, and Mental Health in Afghanistan," 611–38. On medical disability, see Hammen, "Stress and Depression," 293–319. On life-threatening illness, see Elie G. Karam et al., "Cumulative Traumas and Risk Thresholds: 12-Month PTSD in the World Mental Health (WMH) Surveys," *Depression and Anxiety* 31, no. 2 (2014): 130–42.

227 **accident**: Karam et al., "Cumulative Traumas and Risk Thresholds," 130–42.

227 **disability or chronic illness**: On disability, see Hammen, "Stress and Depression," 293–319. On chronic illness, see Bonanno et al., "What Predicts Psychological Resilience after Disaster?," 671. On chronic disease, see Ljiljana Trtica Majnarić et al., "Low Psychological Resilience in Older Individuals: An Association with Increased Inflammation, Oxidative Stress and the Presence of Chronic Medical Conditions," *International Journal of Molecular Sciences* 22, no. 16 (2021): 8970.

227 **victim of a crime**: Fernandez et al., "Assessing the Relationship between Psychosocial Stressors and Psychiatric Resilience among Chilean Disaster Survivors," 630–37.

227 **Abuse**: Karam et al., "Cumulative Traumas and Risk Thresholds," 130–42.

227 **assault**: McLaughlin et al., "Childhood Adversity, Adult Stressful Life Events, and Risk of Past-Year Psychiatric Disorder," 1647–58; Jesse R. Cougle, Heidi Resnick, and Dean G. Kilpatrick, "Does Prior Exposure to Interpersonal Violence Increase Risk of PTSD Following Subsequent Exposure?," *Behaviour Research and Therapy* 47, no. 12 (2009): 1012–17.

227 **Witnessing assault or violence**: Fernandez et al., "Assessing the Relationship between Psychosocial Stressors and Psychiatric Resilience among Chilean Disaster Survivors," 630–37; Cougle, Resnick, and Kilpatrick, "Does Prior Exposure to Interpersonal Violence Increase Risk of PTSD Following Subsequent Exposure?," 1012–17.

227 **The death or illness of a loved one**: Karam et al., "Cumulative Traumas and Risk Thresholds," 130–42; Fernandez et al., "Assessing the Relationship between Psychosocial Stressors and Psychiatric Resilience among Chilean Disaster Survivors," 630–37.

227 **Parental coldness or neglect**: Bart P. F. Rutten et al., "Resilience in Mental Health: Linking Psychological and Neurobiological Perspectives," *Acta Psychiatrica Scandinavica* 128, no. 1 (2013): 3–20.

227 **Parental divorce or death**: McLaughlin et al., "Childhood Adversity, Adult Stressful Life Events, and Risk of Past-Year Psychiatric Disorder," 1647–58.

227 **Parental mental illness or substance abuse**: Corina Benjet, Guilherme Borges, and María Elena Medina-Mora, "Chronic Childhood Adversity and Onset of Psychopathology during Three Life Stages: Childhood, Adolescence and Adulthood," *Journal of Psychiatric Research* 44, no. 11 (2010): 732–40; Cyleen A. Morgan et al., "Adverse Childhood Experiences Are Associated with Reduced Psychological Resilience in Youth: A Systematic Review and Meta-Analysis," *Children* 9, no. 1 (2021): 27.

227 **Bullying or mistreatment**: Heather A. Turner, David Finkelhor, and Richard Ormrod, "Poly-Victimization in a National Sample of Children and Youth," *American Journal of Preventive Medicine* 38, no. 3 (2010): 323–30.

227 **Economic hardship**: Erica M. Webster, "The Impact of Adverse Childhood Experiences on Health and Development in Young Children," *Global Pediatric Health* 9 (2022): 2333794 X221078708.

227 **health issues or illness**: Hamideh Mahdiani and Michael Ungar, "The Dark Side of Resilience," *Adversity and Resilience Science* 2, no. 3 (2021): 147–55; Benjet, Borges, and Medina-Mora, "Chronic Childhood Adversity and Onset of Psychopathology during Three Life Stages," 732–40.

227 **Physical or sexual abuse**: Karam et al., "Cumulative Traumas and Risk Thresholds," 130–42.

227 **violence or victimization**: McLaughlin et al., "Childhood Adversity, Adult Stressful Life Events, and Risk of Past-Year Psychiatric Disorder," 1647–58; Turner, Finkelhor, and Ormrod, "Poly-Victimization in a National Sample of Children and Youth," 323–30.

Appendix C

229 **We are ineffective**: Jungwoo Ha, "Not Being Able to Verify One's Confidence: Negative Consequences of Thwarted Self-Promotion," *Academy of Management Proceedings* 2017, no. 1 (2017).

229 **Worthlessness**: Jungwoo Ha, "The Impact of Thwarted Competence-Presentation on Turnover Intentions," *Academy of Management Proceedings* 2015, no. 1 (2015): https://doi.org/10.5465/ambpp.2015.14824abstract; Hui Fang et al., "Being Eager to Prove Oneself: U-Shaped Relationship between Confidence Frustration and Intrinsic Motivation in Another Activity," *Frontiers in Psychology* 8 (2017): 2123; N. Pontus Leander and Tanya L. Chartrand, "On Thwarted Goals and Displaced Aggression: A Compensatory Competence Model," *Journal of Experimental Social Psychology* 72 (2017): 88–100.

229 **Am I useless**: Kimberley J. Bartholomew et al., "Psychological Need Thwarting in the Sport Context: Assessing the Darker Side of Athletic Experience," *Journal of Sport and Exercise Psychology* 33, no. 1 (2011): 75–102.

229 **Am I no good?**: Marie-Christine Opdenakker, "Need-Supportive and Need-Thwarting Teacher Behavior: Their Importance to Boys' and Girls' Academic Engagement and Procrastination Behavior," *Frontiers in Psychology* 12 (2021): 628064; Evangelos Brisimis et al., "Exploring the Relationships of Autonomy-Supportive Climate, Psychological Need Satisfaction and Thwarting with Students' Self-Talk in Physical Education," *Journal of Education, Society and Behavioural Science* 33, no. 11 (2020): 112–22.

229 **helpless**: Delrue et al., "Game-to-Game Investigation of the Relation between Need-Supportive and Need-Thwarting Coaching and Moral Behavior in Soccer," 1–10.

229 **micromanaged**: Nikita Bhavsar et al., "Conceptualizing and Testing a New Tripartite Measure of Coach Interpersonal Behaviors," *Psychology of Sport and Exercise* 44 (2019): 107–20.

229 **left out or excluded**: Delrue et al., "Game-to-Game Investigation of the Relation between Need-Supportive and Need-Thwarting Coaching and Moral Behavior In Soccer," 1–10.

229 **letting us down**: John Tooby and Leda Cosmides, "Friendship and the Banker's Paradox: Other Pathways to the Evolution of Adaptations for Altruism," *Proceedings-British Academy* 88 (1996): 119–43.

229 **worthy of love**: Kipling D. Williams, "Ostracism: A Temporal Need-Threat Model," *Advances in Experimental Social Psychology* 41 (2009): 275–314.

Appendix G

238 **effective and capable**: Kennon M. Sheldon et al., "Persistent Pursuit of Need-Satisfying Goals Leads to Increased Happiness: A 6-Month Experimental Longitudinal Study," *Motivation and Emotion* 34 (2010): 39–48.

238 **things I am good at**: Sheldon et al., "Persistent Pursuit of Need-Satisfying Goals Leads to Increased Happiness," 39–48.

238 **confident in my actions**: Edward L. Deci and Richard M. Ryan, "Self-Determination Research: Reflections and Future Directions," in *Handbook of Self-Determination Research*, ed. Edward L. Deci and Richard M. Ryan (New York: University of Rochester Press, 2002).

238 **sense of accomplishment**: Kennon M. Sheldon and Jonathan C. Hilpert, "The Balanced Measure of Psychological Needs (BMPN) Scale: An Alternative Domain General Measure of Need Satisfaction," *Motivation and Emotion* 36 (2012): 439–51.

238 **overcome obstacles**: Richard M. Ryan, Veronika Huta, and Edward L. Deci, "Living Well: A Self-Determination Theory Perspective on Eudaimonia," *Journal of Happiness Studies* 9 (2008): 139–70.

238 **interesting new skills; good at what I do**: Sheldon and Hilpert, "Balanced Measure of Psychological Needs (BMPN) Scale," 439–51.

238 **made to feel ineffective; make me feel incompetent**: Sebastiano Costa, Nikos Ntoumanis, and Kimberley J. Bartholomew, "Predicting the Brighter and Darker Sides of Interpersonal Relationships: Does Psychological Need Thwarting Matter?," *Motivation and Emotion* 39 (2015): 11–24.

238 **unrealistic expectations of me, feel inferior, fulfill my potential**: Kimberley J. Bartholomew et al., "Self-Determination Theory and Diminished Functioning: The Role of Interpersonal Control and Psychological Need Thwarting," *Personality and Social Psychology Bulletin* 37, no. 11 (2011): 1459–73; Jochen Delrue et al., "A Game-to-Game Investigation of the Relation between Need-Supportive and Need-Thwarting Coaching and Moral Behavior In Soccer," *Psychology of Sport and Exercise* 31 (2017): 1–10.

239 **Other people doubt**: Marie-Christine Opdenakker, "Need-Supportive and Need-Thwarting Teacher Behavior: Their Importance to Boys' and Girls' Academic Engagement and Procrastination Behavior," *Frontiers in Psychology* 12 (2021): 628064.

239 **like a failure**: Opdenakker, "Need-Supportive and Need-Thwarting Teacher Behavior," 628064.

Appendix H

240 **authentic interests and values**: Kennon M. Sheldon and Jonathan C. Hilpert, "The Balanced Measure of Psychological Needs (BMPN) Scale: An Alternative Domain General Measure of Need Satisfaction," *Motivation and Emotion* 36 (2012): 439–51.

240 **make my own choices**: Meredith Rocchi et al., "Assessing Need-Supportive and Need-Thwarting Interpersonal Behaviours: The Interpersonal Behaviours Questionnaire (IBQ)," *Personality and Individual Differences* 104 (2017): 423–33.

240 **how to life my life; express my ideas and opinions; pretty much be myself**: Sheldon and Hilpert, "Balanced Measure of Psychological Needs (BMPN) Scale," 439–51.

240 **express my true self**: Pedro Cordeiro et al., "The Portuguese Validation of the Basic Psychological Need Satisfaction and Frustration Scale: Concurrent and Longitudinal Relations to Well-Being and Ill-Being," *Psychologica Belgica* 56, no. 3 (2016): 193.

Notes

240 **choose for myself**: Sebastiano Costa, Nikos Ntoumanis, and Kimberley J. Bartholomew, "Predicting the Brighter and Darker Sides of Interpersonal Relationships: Does Psychological Need Thwarting Matter?," *Motivation and Emotion* 39 (2015): 11–24.

240 **unwanted pressure**: Tiphaine Huyghebaert-Zouaghi et al., "Advancing the Conceptualization and Measurement of Psychological Need States: A 3 × 3 Model Based on Self-Determination Theory," *Journal of Career Assessment* 29, no. 3 (2021): 396–421; Sheldon and Hilpert, "Balanced Measure of Psychological Needs (BMPN) Scale," 439–51.

240 **external demands**: Marie-Christine Opdenakker, "Need-Supportive and Need-Thwarting Teacher Behavior: Their Importance to Boys' and Girls' Academic Engagement and Procrastination Behavior," *Frontiers in Psychology* 12 (2021): 628064.

240 **I feel controlled**: Evangelos Brisimis et al., "Exploring the Relationships of Autonomy-Supportive Climate, Psychological Need Satisfaction and Thwarting with Students' Self-Talk in Physical Education," *Journal of Education, Society and Behavioural Science* 33, no. 11 (2020): 112–22.

240 **behave in certain ways**: Jochen Delrue et al., "A Game-to-Game Investigation of the Relation between Need-Supportive and Need-Thwarting Coaching and Moral Behavior In Soccer," *Psychology of Sport and Exercise* 31 (2017): 1–10.

240 **pushed to do things**: Kimberley J. Bartholomew et al., "Self-Determination Theory and Diminished Functioning: The Role of Interpersonal Control and Psychological Need Thwarting," *Personality and Social Psychology Bulletin* 37, no. 11 (2011): 1459–73.

240 **prevented from making choices**: Costa, Ntoumanis, and Bartholomew, "Predicting the Brighter and Darker Sides of Interpersonal Relationships," 11–24.

240 **what I have to do**: Sheldon and Hilpert, "Balanced Measure of Psychological Needs (BMPN) Scale," 439–51.

241 **follow decisions or plans**: Bartholomew et al., "Self-Determination Theory and Diminished Functioning," 1459–73; Costa, Ntoumanis, and Bartholomew, "Predicting the Brighter and Darker Sides of Interpersonal Relationships," 11–24.

241 **because I have to**: Sheldon and Hilpert, "Balanced Measure of Psychological Needs (BMPN) Scale," 439–51.

Appendix I

242 **Are You Overidentified at Work?**: This scale is inspired by an excellent article by Janna Koretz, "What Happens When Your Career Becomes Your Whole Identity," *Harvard Business Review*, December 26, 2019, https://hbr.org/2019/12/what-happens-when-your-career-becomes-your-whole-identity.

Appendix J

244 **how this person wronged me**: Nathaniel G. Wade et al., "Measuring State-Specific Rumination: Development of the Rumination about an Interpersonal Offense Scale," *Journal of Counseling Psychology* 55, no. 3 (2008): 419.

Appendix K

246 **pretty friendly to me**: Sebastiano Costa, Nikos Ntoumanis, and Kimberley J. Bartholomew, "Predicting the Brighter and Darker Sides of Interpersonal Relationships: Does Psychological Need Thwarting Matter?," *Motivation and Emotion* 39 (2015): 11–24.

Notes

246 **regularly interact with to be my friends**: John Tooby and Leda Cosmides, "Friendship and the Banker's Paradox: Other Pathways to the Evolution of Adaptations for Altruism," *Proceedings-British Academy* 88 (1996): 119–43.

246 **usually relate well to me**: Meredith Rocchi et al., "Assessing Need-Supportive and Need-Thwarting Interpersonal Behaviours: The Interpersonal Behaviours Questionnaire (IBQ)," *Personality and Individual Differences* 104 (2017): 423–33.

246 **intimacy and trust**: Arlen C. Moller, Edward L. Deci, and Andrew J. Elliot, "Person-Level Relatedness and the Incremental Value of Relating," *Personality and Social Psychology Bulletin* 36, no. 6 (2010): 754–67.

246 **network that encourages me**: Ellen Skinner and Kathleen Edge, "Self-Determination, Coping, and Development," in *Handbook of Self-Determination Research*, ed. Edward L. Deci and Richard M. Ryan (Rochester, NY: University of Rochester Press, 2002), 297–337.

246 **give back to me**: Nele Laporte et al., "Adolescents as Active Managers of Their Own Psychological Needs: The Role of Psychological Need Crafting in Adolescents' Mental Health," *Journal of Adolescence* 88 (2021): 67–83.

246 **one close and true friend**: Roy F. Baumeister and Mark R. Leary, "The Need to Belong: Desire for Interpersonal Attachments as a Fundamental Human Motivation," *Interpersonal Development* (2017): 57–89.

246 **do not seem to like me**: Costa, Ntoumanis, and Bartholomew, "Predicting the Brighter and Darker Sides of Interpersonal Relationships," 11–24.

246 **dismissive of me**: Kimberley J. Bartholomew et al., "Self-Determination Theory and Diminished Functioning: The Role of Interpersonal Control and Psychological Need Thwarting," *Personality and Social Psychology Bulletin* 37, no. 11 (2011): 1459–73.

246 **don't fit in**: Marie-Christine Opdenakker, "Need-Supportive and Need-Thwarting Teacher Behavior: Their Importance to Boys' and Girls' Academic Engagement and Procrastination Behavior," *Frontiers in Psychology* 12 (2021): 628064.

247 **keep to myself**: Costa, Ntoumanis, and Bartholomew, "Predicting the Brighter and Darker Sides of Interpersonal Relationships," 11–24.

247 **Important people in my life don't comfort me**: Rocchi et al., "Assessing Need-Supportive and Need-Thwarting Interpersonal Behaviours," 423–33.

247 **let me down**: Tooby and Cosmides, "Friendship and the Banker's Paradox," 119–43.

Index

Page numbers followed by a t *indicate a table.*

acceptance and commitment therapy (ACT), 85
adrenaline, 19, 21
Aniston, Jennifer, 32
anxiety, 34, 38, 54–55, 111
anxiety attacks, 54–55, 57
The Art of Resilience (Edgely), 32
asking for help, 56–57
authenticity, 57, 162–63
autoimmune disease, 128–32, 170, 177, 214, 217
autonomy support, 166–69
Avalon House, 204–5
Avalon Village, 205–6
avoidance. *See* emotion avoidance

bad guys bias, 196, 201
bad things bias, 20–21, 25
belonging, 185
betrayal, 188
betrayal trauma, 125n5
better way mindset, 65, 67
black-and-white thinking, 151, 156
brain fog, 43–44
breaking point, 3–6. *See also* resilience ceilings

Broadway musicals, 11, 111–12, 185n3, 215–16
Brown, Brené, 79, 192
Buffy the Vampire Slayer, 15, 212–13
bully jujitsu, 164–66, 178
burnout, 47
Byron, Lord, 77

Calhoun, Lawrence, 60–61
Cameron, Jo, 78
Camus, Albert, 200
Carrell, Steve, 137
chaos
    as confidence trigger, 139
    era of, 18–20, 25
    harnessing, 8
Chen, Nathan, 99–100, 111–13, 139
choice
    activities supporting, 179t
    authenticity and, 162–63, 179–80t
    bully jujitsu (humor) and, 164–66, 178
    choice support and, 166–69, 180
    crafting choice in one category and, 171–73, 178

choice (*cont.*)
  overidentification and, 173–75, 178
  public sector employees and, 162n5
  recovering authentic identity and, 175–76, 178–79
  reducing uncertainty and, 167–68
  shadow goals/habits associated with, 117*t*, 163–64, 178
  triggers from thwarted, 10, 93, 103, 162–63
  2-2-2 tool, 169–70, 178
Cohen, Leonard, 60
compensatory motives, 109–10
conditional acceptance, 151, 156
confidence
  activities supporting, 179*t*
  competence wins, 149–50
  confidence triggers, 103, 139–40
  correlation to ability and, 138–39
  elements of, 138
  external challenges and, 153–54, 157
  future you exercise, 148–49
  impostor syndrome and, 140–41, 144
  metaperception and, 145
  perfectionism and, 142–44
  reflected best self exercise (RBS), 146–47, 156
  setbacks and, 154–57
  shadow goals/habits associated with, 117*t*, 156–57*t*
  as source of trigger, 10, 92
  10 percent buffer approach and, 150–52, 157
conflict, 187–88
connection
  activities supporting, 179*t*
  backers vs. barnacles and, 190–91, 201
  bad guys bias, 196, 201
  belonging and, 185, 201
  building blocks of, 184–85, 201
  creative perspective taking, 197, 201
  declining social connections and, 186–87
  disconnecting from past and, 191–93
  loneliness vs., 10, 184–87, 200
  nostalgia and, 182n1, 201
  offense rumination and, 195–96, 201
  pair bonds and, 189–90, 201
  relationship depth and, 185, 201
  shadow goals connected to thwarted, 189–90
  shadow habits connected to, 202*t*
  spirituality and, 198–201
  transforming conflict, 194–98
  triggers connected to, 93, 187–89, 200
coping capacity, 47–48
cortisol, 21–22, 25
costly persistence, 51, 53
Counterfeit Gods (Keller), 198
COVID lockdowns, 34, 121

Index

creative perspective taking, 197, 201
crisis, 60–61
cruelty, 188
Curran, Thomas, 143

de Berker, Archy, 22
Deci, Edward, 91–92
DeGeneres, Ellen, 205
depression, 34
dichlorodiphenyltrichloroethane (DDT), 16
Dinner for Schmucks, 137–38
disassociation, 76. *See also* emotion avoidance
divorce, 55, 125, 192n7
Duckworth, Angela, 46

Edgely, Ross, 32
Ehlers-Danlos Syndrome (EDS), 129–31, 170, 177, 214, 217
Elsayed, Nabeela, 52, 54–58, 62–65
emotion avoidance
   assessment, 88
   emotion suppression and, 72–74, 86
   freeze-or-faint system, 75–77, 86
   tools to overcome, 83–85
   toxic positivity, 74–75, 86
emotion suppression, 72–74, 86
expectations, 139
exploration network, 196, 201
exploratory dialog, 196–97, 201
extrinsic motivation, 111, 116
eye movement desensitization and reprocessing (EMDR), 98n9

Faulkner, William, 97
fight-or-flight response, 18, 75
financial goals, 126n6
Ford Motor Company, 203
forgiveness, 83–84, 86
freeze-or-faint system, 75–77, 83, 86
Freud, Sigmund, 107
future you exercise, 148–49, 157

Geher, Glenn, 188
Goethe, Johann Wolfgang von, 106
Goldsmith, Marshall, 63, 216
Goodman, Whitney, 74
grit gaslighting, 9, 38, 42, 71, 80
Groban, Josh, 138–39
grow forward plan, 64

Hadestown, 111
Harris, Shamayim, 203–5
Hayes, Stephen, 210
Heath, Dan, 94
Hill, Andrew, 143
Hold on Tight, 81
Holling, Crawford Stanley "Buzz." 29–30, 32–33, 60
Hone, Lucy, 20
"How Did Healing Ourselves Get So Exhausting?," 51
humor, 164–66

Ikea, 55–57
impostor syndrome, 140–41, 144, 156
inertia trap, 210
inflammation, 17

Index

integrative emotion regulation, 79n7
International Criminal Court, 169
intrinsic motivation, 111, 116, 124, 146

Jung, Carl, 106–8, 115, 213

Keller, Tim, 198
Kennedy, John F., 60
kintsugi, 59–60
Kostunica, Vojislav, 168, 169

Laporte, Nele, 121
leaves on a stream technique, 85–86
life crafting, 208

Mair, Victor, 60
mantras, 129
Maslach, Christina, 47
mast cell activation syndrome, 130–31, 177
metaperception, 145, 156
Milošević, Slobodan, 159–61, 168–69
Miserandino, Christine, 48
monotony, 139
mood release, 84–86
Mulally, Alan, 65
Muraven, Mark, 40
Musk, Elon, 32
mustivation, 17

natural disasters, 16
need audit, 102–5
need crafting, 120–22, 132, 207

negativity bias, 20–21, 25
negativity rebounds, 73
neglect, 187
Nesse, Randolph, 78
Nietzsche, Friedrich, 39, 41
noradrenaline, 21

Obama, Michelle, 32
Odum, Eugene, 30
offense rumination, 195–96, 201
100 Coaches Program, 216
Otpor!, 161, 164–69
overload trap, 209

pain
  avoidance of, 72–77, 86
  befriending, 82–83, 86
  as data, 81–83
  as purpose, 79–81
  as signal, 77–79
  as source of power, 79, 86
pain paradox, 72–74
pair bonds, 189–90
Park, Crystal, 60–61
people pleasing, 75
perfectionism, 142–44
Perkins, George, 30
persistence, costs of, 51, 53
personal narrative, taking charge of, 84, 86
physical symptoms (from stress), 17, 24n6, 51, 142
pivoting
  meaning of, 119, 132
  sentinel events and, 119–20, 132
  shatterproof habits, 124–27, 132

polyvagal theory, 75–77, 83, 86
  tools for, 83–85
Popović, Srdja, 160–61, 164–67
Porges, Stephen, 76, 83–84
post-traumatic stress disorder (PTSD), 95, 125n5
postural orthostatic tachycardia syndrome, 130, 177
protective factors, 28, 31–32, 34
Putnam, Robert, 187

Rauschenbach, Emma, 107
reflected best self exercise (RBS), 146–47, 156
rejection, 187
relational trauma, 125n5
relationship depth, 185
resilience
  burnout vs., 47
  costs of, 51, 53
  cultural appreciation for, 31
  depletion of, 39–40, 53
  discrimination and, 40–41
  dos and don'ts, 53
  of ecosystems, 30
  overreliance on, 48
  predictive value of, 35–36
  shifting definitions of, 32–33
  skin-deep, 49–51, 53
Resilience Alliance, 30
resilience ceiling
  hitting, 3–6, 9, 46–47, 52–53
  indications of, 49, 53
  quiz, 49, 53, 211
  spoon theory and, 48, 53
The Resilience Factor, 32

resilience myths
  better-and-stronger myth, 34–36, 42
  resilience-as-choice myth, 36–39, 42
  what-doesn't-kill-us myth, 7, 39–42, 61
resilience research
  interventions and, 37–38
  practices for resilience, 28, 32, 35–37, 42
  protective factors, 28, 31–32, 34
  risk factors and, 38–39
resilience study (author's)
  bad things and, 23–25
  breaking points and, 6–7
  need crafting and, 122
  physical symptoms of stress and, 24n6
  resilience scale and, 33
reverse compass, 210
Rimtutituki, 164
Rising Strong (Brown), 79
Rudd, Paul, 137
Rumi, 60
Rutter, Michael, 46
Ryan, Richard, 91–92

Schofield, Jill, 131, 176–77, 210, 214
Schulkin, Jay, 78
self-awareness, 209
self-determination theory (SDT), 91–92, 103, 120
sentinel events, 119–20, 125, 132, 190

shadow goals, 109–13, 116
  common, 177t, 179–80t
  compensatory motives, 109–10
  confidence triggers and, 156–57t
  prevention and, 110, 116
  proof and, 110, 116
  self-protection, 109–10, 116
  sources of, 110–13, 116
shadow habits, about, 113–16
  choice triggers and, 179–80t
  confidence triggers and, 139–40, 156–57t
shadows
  Jung's work on, 107–8, 116
  shadow goals., 109–10, 116, 117t
  shadow habits, 113–16
  sources of shadow goals and, 110–13
"Shake It Off" (Swift), 32
shame, hitting resilience ceiling and, 47
shatterproof habits, strategic experiments and, 124–27, 132
shatterproof mindset
  becoming focus and, 61, 67
  change and, 63, 67
  embracing pain and, 62–63, 67
  growing forward and, 64–65, 67
  need fulfillment and, 61–62
  as proactive, 61
  resistance mindset vs., 61
  shatterproof practices, 62
Shatterproof Road Map
  arguments for using, 65–66
  insights for, 207–9

  overview of, 8–11
  practices, 133t
  traps to avoid, 209–11
Shatterproof Six, 120–22, 123t, 124, 132
Singh, Simran Jeet, 40
skin-deep resilience, 49–51, 53
sleep disturbances, 17, 143
social trust, 187
Sondheim, Stephen, 11
spirituality, 198–201
spoon theory, 48
Spring, Carolyn, 95
spruce budworm, 15–16, 29–30
Stoicism, 31, 72–73
stress
  resilience depletion and, 39–40
  symptoms of, 17, 24n6, 51, 142
stressed-out strivers, 7, 16–17, 25
The Sun Also Rises (Hemingway), 90–91
Szostak, Stephanie, 137–38

Tasha Ten program, 216
Tedeschi, Richard, 60–61
10 percent buffer, 150–52, 157
Thompson, Derek, 186–87
three-minute mood map, 85–87
three-to-thrive needs, 92–94, 103
  need crafting, 120–22
  need audit, 102–5
  sentinel events and, 120
  See also choice; confidence; connection
Thurston, Baratunde, 49–51

Index

tools
    creative perspective taking, 197
    forgiveness, 83–84, 86
    future you exercise, 148–49, 157
    leaves on a stream technique, 85–86
    mood release, 84–85, 86
    need audit, 102–5
    reflected best self exercise (RBS), 146–47, 156
    resilience ceiling quiz, 49, 53, 211
    reverse compass, 210
    taking charge of narrative, 84, 86
    10 percent buffer, 150–52, 157
    three-minute mood map, 85–87
    2-2-2 tool, 169–70, 178
toxic positivity, 74–75, 86
trauma
    post-traumatic growth and, 60–61
    professional care and, 11
    Serbia under Milošević and, 164
triggers
    identifying patterns in, 100–101, 103
    post-traumatic stress disorder (PTSD) and, 95
    self-determination theory (SDT) and, 91–92, 103
    as signals of unmet needs and, 95–98, 103
    three-to-thrive needs and, 92–94, 103
    tracing to unmet needs, 101–2, 103
2008 financial crisis, 32
2-2-2 tool, 169–70, 178

uncertainty, 22–23, 25, 167
Ungar, Michael, 37
University Act (Serbia), 161
Upstream (Heath), 94

Walmart Canada, 57
Wang, Hetty, 99
Werner, Emmy, 31, 51
Wilcox, Ella Wheeler, 176
Wiseman, Liz, 47
workers
    burnout vs. resilience and, 47
    environmental stress and, 16–17, 25
    stressed-out strivers, 7, 16–17, 25

Yoshimasa, Ashikaga, 59

About the Author

Dr. Tasha Eurich is an organizational psychologist, researcher, and *New York Times* bestselling author on a mission to help people and organizations thrive in an era of constant change.

Recognized as the world's leading self-awareness coach (Marshall Goldsmith Coaching Awards) and communication expert (Global Gurus), Eurich has spent over two decades teaching people practical strategies to supercharge their self-awareness, sanity, and success. She's worked directly with over 40,000 leaders—and spoken to hundreds of thousands more—on every continent but Antarctica.* Her 2017 TEDx talk has been viewed more than 10 million times.

With a PhD in Industrial-Organizational Psychology, Eurich has been named one of the world's most influential coaches by Thinkers50 and Coaching.com. As principal of The Eurich Group, she is trusted by some of the world's most powerful leaders—from Fortune 500 CEOs and founders to the occasional NBA coach. Her clients include Google, Walmart, Salesforce, Nestlé, T-Mobile, Royal Bank of Canada, Johnson & Johnson, and the White House Leadership Development Program.

Eurich's work has been globally recognized for its blend of powerful lessons and practical tools. Her first book, *Bankable Leadership*, was hailed by *Shark Tank*'s Barbara Corcoran as "a refreshing approach that can change both lives and businesses." Her second, *Insight*, was named the #1 career book by The Muse, sits on Brené Brown's bookshelf, and is one of the three books famed Wharton professor Adam Grant recommends most often. She has been featured in outlets like the *Wall Street*

---

*Does anyone know anyone in Antarctica?

Journal, the *New York Times,* the *Harvard Business Review,* NPR, CNN, NBC, and *Fast Company,* as well as peer-reviewed journals.

As a proud member of Marshall Goldsmith's 100 Coaches, Eurich was the first to launch her own pay-it-forward program, The Tasha Ten. Its purpose is to build a global community of leaders who embody and champion human-centered leadership, driving positive change in businesses, institutions, and communities worldwide. Its thirteen active members and five emeritus members represent eleven countries around the world.

In her spare time, she enjoys traveling and rescuing dogs, and she is a proud and unapologetic musical theater nerd.